WE SHALL TRAVEL IN SPACE-TIME

WE SHALL TRAVEL IN SPACE-TIME

MEMORY OF AUTHOR CRITICAL STUDIES ON
RELATIVITY THEORY, SPACE-TIME TRAVELS
AND WORLD FRACTAL STRUCTURE

VALLEC ORCIANAR

Copyright © 2020 by Vallec Orcianar.

Library of Congress Control Number:		2020903425
ISBN:	Hardcover	978-1-9845-9392-4
	Softcover	978-1-9845-9393-1
	eBook	978-1-9845-9394-8

All rights reserved. No part of this book may be reproduced or transmitted in any form or by any means, electronic or mechanical, including photocopying, recording, or by any information storage and retrieval system, without permission in writing from the copyright owner.

Any people depicted in stock imagery provided by Getty Images are models, and such images are being used for illustrative purposes only.
Certain stock imagery © Getty Images.

Print information available on the last page.

Rev. date: 02/20/2020

To order additional copies of this book, contact:
Xlibris
800-056-3182
www.Xlibrispublishing.co.uk
Orders@Xlibrispublishing.co.uk

CONTENTS

Introduction ix

Chapter 1   The Constant of Light's Speed the Measurements 1
Chapter 2   The Transformation Factor 5
Chapter 3   Michelson–Morley Experiment 9
Chapter 4   The Mathematical Way to Obtain the LF 15
Chapter 5   The General Transformation Factor 21
Chapter 6   The Transformations From the Factors of Transformations 25
Chapter 7   Relativistic Dynamic 29
Chapter 8   Mass and Speed 35
Chapter 9   Thomson's Experiment 41
Chapter 10 The Rogers, Reynolds, Rogers Experiment 43
Chapter 11 Oscillating Speed Energy 47
Chapter 12 Fossil Radiation Uniformity 51
Chapter 13 Fizeau–Zeemann Experiment 55
Chapter 14 Black Holes Evaporation 59
Chapter 15 Matter Formation 63
Chapter 16 Light Refraction 69
Chapter 17 How to Travel in Space–Time 73
Chapter 18 Space–Time Travel Projecting 75
Chapter 19 Travelling Towards the Future 77
Chapter 20 Back in Time Travels 83
Chapter 21 Conclusions 87
Chapter 22 Ballistics 91
Chapter 23 Epilogue 97

Dedicated
to
TILDE
and
VICO
my
unforgettable
parents

Introduction

Space–time might be defined, in Albert Einstein's words, as the 'continuum' of space and time on which we are and move. Here is relevant the idea of continuity from space and time which appears as consequence of the motion between coordinates systems (bodies). The study of what happens to space and time by effect of motion of the bodies gives rise to a relativity theory that Einstein qualified as 'restricted' or 'special' when the motion between the systems should be at constant speed.

This theory is founded on the remarkable mysterious fact of the constant of light's speed.

But the 'restricted relativity', so as given to the press in the 1905, would never permit to travel in time, because, to do it, it would be necessary to reach the speed of the light, but to arrive to such, by Einstein's relativistic dynamic, should be necessary an infinite energy.

But Einstein was wrong. This is what this memory must show, as mean to show what is objectively important; it is the possibility to travel in time, arriving, as consequence, to create a new relativity theory starting from the foundations. As already said, the foundation of a relativity theory is the constant of light's speed. But this does not mean that the speed of light does not vary with time, which is an old well-consolidated physics idea, but makes reference to the new (at the end of nineteenth century) discovery that light does not follow the laws of Galileo's relativity and then, for instance, a light's beam emitted by a system driving away from, or approaching to another one, will be measured of the same speed by both the systems. A speed enough accurately near the 300,000 km/s. We will assume such value for the calculations.

But after having cleared that if it is true that each system, when they are in relative motion measures the light's speed always equal to 300,000 km/s, must be said that, for a detached observer those speeds are not equal. From this come the problems that a relativity theory must solve, of different importance and value, depending by different view points. The privileged view point of this memory is that of time travels problems. Others problems faced Einstein, driving him to conclusions some time denying the chance of time travels. In those cases, a criticism to show the supposed mistakes of Einstein's relativity will be inevitable.

What can be anticipated on this issue is that Einstein's relativity is a wrong answer to an irrelevant question. The question is this: 'Given two bodies in relative motion with constant speed in one direction, which modifications would affect the bodies?' The relevant question would be this: 'Given the same bodies, which modifications would affect space and time all around them?' But Einstein's relativity is wrong too, as will be shown more ahead, and the correction of the mistakes will open interesting perspectives on the chance to travel on time.

Another remark must be made, this referring to exposition method. In the memory will be faced several problems, not taking into count the wonderful means that modern mathematics puts at disposition. This is not just a choice of shortcuts, but of simplifications, necessary to meet the average cultural level of the readers.

To this purpose have been introduced very concrete explanations of phenomena so to escape to quantum physics abstraction as to avoid analogies often inappropriately used in divulgation works. The followed criterion is then to introduce easily understandable hypothesis rather than difficult theories that would have to be dogmatically accepted.

Will not be avoided anyway by means of easy hypothesis, the hunt to the mistakes of difficult theories. It is possible to cite here the example of the research of the transformation factor for the special relativity that, drown by Einstein by means of *linear algebra* arrives to the Lorentz factor (LF), which will be shown inadequate to the case by a more elementary mathematical means.

The same may be said of the means Einstein followed to arrive at his very famous relation $E = mC^2$. He adopted a mathematical proceeding without relation with the physical meaning of the formula, when there are at least two hypotheses by easily understandable meaning that drive to the same $E = mC^2$. They will be presented ahead.

But in these pages have been brought several other proofs of the validity of the relativity theory here sustained; they go from a new interpretation of the working of charged particle accelerators (to explain why the particles never reach the light speed) to the explication, more quantitatively accurate than the standard one, of the long passage of the mesons in the atmosphere; from the explication of the why of the uniformity of the fossil radiation (why all that radiation arrive on Earth with the same temperature when it comes from points so differently far, so that the emission must necessarily have taken place at very different moments and then at very different temperature, to the right explication of the result of the *Fizeau–Zeeman* experiment, which the Einsteinian relativity pretend to explain recurring to the speed's relativistic composition, but mistakenly for two reasons: in the influence of the transformation factor and the determination of light's speed in water in motion.

And how might evaporate the black holes, according to the thesis of Stephen Hawking if the same light is, by them, captured? The relativity here supported can explain it, so as explaining the oscillating nature of photons and the origin of positive and negative electrical charges; which drives so to unification of classical with quantum physics, as to unification of gravitational and electromagnetic interactions.

Note 1: Some of the over made speeches require the introduction of an elementary particle today well known as 'Higgs boson' because of the researches in act in Geneva at the *Large Hadron Collider* to make it appear. To underline its absolute elementarity (nothing smaller and all the matter is, at last, made of it) it has been also called 'God's particle'. Not having a well-defined established name, in this memory it will be called Absolutely Elementary Particle (AEP).

At this point it appears logical that the photons too be made of AEP. They must then be thought as waves of AEP. The fact of they having moment, proves their corpuscular nature.

Note 2: In some cases, in this memory, the word 'tensor' will be used without respect for the acquired mathematical meaning. Will be done this operation because and when the word will well express the time–space deformations recalling the idea of tensions.

Chapter 1

The Constant of Light's Speed the Measurements

The light's speed constant is the fundamental discovery for the formulation of a relativity theory. It consists in the ascertainment that the light *moves* in the void, with the same speed $C = 300,000$ km/s independently by the relative motion of coordinates' systems that emit or receive it.

Before proceeding, it is convenient to reflect on the underlining of the word *moves*, because this is what appeared to the physics of more than a century ago. The light, it is said, appears to possess its own constant speed although it would be more correct to say that it *is measured* with such a constant speed.

The result of the measurements having been identical so at the light's emission point, as at the reception's point, even when the two points were in relative motion so of farthering as of nearing, constituted a problem that affected both the light's nature and the mean in which it moved. At respect, had been made several hypotheses: as nature of the mean had been thought the ether, which could be standing in all the universe being undisturbed by the motion of the bodies in it, but able to vibrate to transmit light's waves, or else had been thought an ether dragged by the bodies moving in it. Other hypothesis: no mean to transmit waves, but completely void space in which the light could move as massless projectiles.

From these fundamental hypotheses, we obtain the following prospect of the expected at the measurements results:

(Ce = Light's emission speed)
(Cr = Light's reception speed)
(v = Systems' relative speed)
(C = Light's absolute speed)

Light as Wave in Still Ether.
(Moving emitter and still receptor)
A) At reception: $Cr = C \pm v$ (− for approaching)
B) At emission: $Ce = C$
(Still emitter and moving receptor)
A) At reception: $Cr = C \pm v$ (+ for approaching)
B) At emission: $Ce = C$

Light as Wave in Dragged Ether
(Moving emitter and standing receptor)
a) At reception: $Cr = C \pm v$ (+ for approaching)
b) At emission: $Ce = C$
(Standing emitter and moving receptor)
a) At reception: $Cr = C$
b) At emission: $Ce = C \pm v$ (+ for approaching)

In this hypothesis the ether is thought completely dragged by the moving point.

Light as Projectiles in Void Space
(Moving emitter and standing receptor)
a) At reception: $Cr = C + v$ (+ for approaching)
b) At emission: $Ce = C$
(Standing emitter and moving receptor)

a) At reception: $Cr = C \pm v$ (+ for approaching)
b) At emission: $Ce = C$

It must be observed that for the physics of the end of nineteenth century and the beginning of the twentieth, the meaning of void space was of absolute, undeformable space – so absolute was the light's speed moving in it. These conceptions had to change just by the trial to solve the measurements' problems, because, differently by the expectations listed in the prospect, those gave always: $Cr = Ce = C$.

It was Einstein who first arrived at the conclusion that, to accept the idea of the light's speed constant, it was necessary to abandon the idea of absolute space and time and to consider them instead as variable depending on the motion of coordinate systems in them. Then not only space but time too is subject to modifications and proportionally to the space, given that the constant of C means the constant of s/f ($= C$) where $s =$ space travelled by a light beam during the time f. But there exists the chance to make confrontations between light's speeds in different coordinates systems and to note they are different from each other. On this base, also times and spaces are confrontable from different coordinates systems. Much more, a relativity theory must just be engaged to find out the means to do this comparison. From these will emerge, for what is here important, not only that in some systems the time will run very slow, and in others very fast, but also that it may run backward too. Conditions and consequences of these premises will be examined in the following pages.

Chapter 2

The Transformation Factor

A restricted relativity theory (the one necessary in this memory) takes into account only coordinate systems possessing straight and uniform speed between them. On this base and on that of the light's speed constant, the systems shall be considered equivalent so that no one can be thought standing, as any one can be chosen as standing relatively to the others. It worths, in short, the reciprocity principle; for, what could be defined of a system respectively to the other, can too be defined of the other respectively to the first.

What will be attempted to define is how spaces, times, and phenomena appear remarked in one system when observed by another system in straight, uniform motion with respect to the first. This draws to find transformation's factors that applied to physical greatnesses of one system give the value of the analogue greatnesses of the other system.

Let's start with the researching the transformation factor for the simplest case – that of a moving system as seen by a standing system and given the coincidence of the direction of the moving system and of the light's beam.

Fig. 1.0

Let's lean the speech on Fig. 1.0 and O are the origins, on x–x axis, of the two systems. Let's consider at first 0 standing and O moving with V speed towards the positive vers of x–x axis. By reciprocity principle, V is seen equal by both the systems. After the time t, measured on O, a light's beam, made start from O origin, at the passage of O by O, reaches the P point, having run the space $X = Ct$, where C = light's speed in O (measured in O). At the same time t, O shall have run the space vt (always measured by O), when, always seen by O, the light's beam shall have run, in O_1, the space $X_1 = C_1 t$. But $X_1 = X - vt$. Then $C_1 t = Ct - vt$; which gives $C_1 = c - v$. Dividing this relation by C at both members, we have $\frac{C_1}{C} = \frac{C}{C} - \frac{V}{C}$ and then $\frac{C_1}{C} = 1 - \frac{V}{C}$ (1). But $\frac{C_1}{C}$ is equal to X_1/X and is given only in function of V and C which are known. Then, that is just the relation that permits to state the O_1 spaces knowing the correspondent ones of O. It is then the valid transformation factor for the considered case.

For the case of considering still O and moving O, we will have $X_1 = Ct$ and $X = X_1 - vt$ as too $X = C_1 t$. Then $C_1 t = Ct - vt$ that drives to the same result as over $C_1/C = 1 - v/e$, proving the reciprocity principle's validity.

Remain to consider the case on which V should be of opposite verse with respect to C. The spaces X and X are now those of the double lines of Fig. 1. We will have a new $X = Ct$ and $X_1 = C_1 t$, but this time $X_1 = X + vt$, at the difference of before. Then $C_1 t = Ct + vt$ and $C_1 = C + V$. Then

$C_1/C = 1 + V/C$ = TF (2). Then the most general transformation factor (TF) for the case of the space in the direction of the motion shall be $C_1/C = 1 \pm V/C$ = TF (3) with the sign − for the case of the systems' motion in accord with that of the light and the + when the sense of motion is opposite.

This means that the motion of a system compresses the space ahead and dilates it behind.

It must be repeated that the TF found for the spaces is valid for the times too, because of the light's speed constant. Indeed we must have always $C = X/t = x \cdot TF/t \cdot TF = X_1/t$.

But the Einsteinian relativity does not agree on several points of the done speech. They are as follows:

1) Different TF
2) For Einstein, modified by the motion is only the same moving body and not differently ahead and behind.
3) For the authors that follow Einstein's relativity, spaces and times modify reversely.
4) For Einstein, time and space are not affected, out of motion's direction.

Here is what can be said of such disagreements:

Disagreement No. 1

For Einstein, the general transformation factor, valid then also for the cases discussed before, would be $LF = 1/\sqrt{1-v^2/c^2}$ and the transformed space: $X_1 = (x-vt)/\sqrt{1-v^2/c^2}$. For us it was $X_1 = X(1 \pm v/c)$.

Our assumption was that $X_1 = X - vt$ is already the space modified by the motion ahead of the moving system. Moreover, Einstein didn't take into count the space modification behind the system that drives to a different ad conjugated $X_1 = X + vt$.

Why Einstein thought to have to find a factor to modify a spatial value, evidently already modified? Unuseful to conjecture, let's go rather to discover the mistakes made by some author trying to justify that factor $\sqrt{1-v^2/c^2}$, the Lorentz factor, named after its discoverer, which needed it to solve electromagnetic problems. Let us start with the most famed Richard P. Feynman, who won the Nobel Prize in Physics in 1965. In his book, *The Feynman Lectures on Physics*, he treats of the LF examining the *Michelson–Morley's* experiment.

Chapter 3

Michelson–Morley Experiment

This experiment by its projector's intentions should have to verify the hypothesis of the existence of ether.

To the purpose was used an apparatus constituted of two perpendicular arms of same length along which were projected two monochromatic light's beams resulting by undoubling of an only beam created at arms' connection point. The undoubling was obtained by interposing O along the original beam of a semitransparent mirror that envoied part of the light along one arm of the apparatus and part along the other arm perpendicular to the first. At the end of the arms, two mirrors reflected the beams to the origin where we collected them by an interferometer which analysed the unfasement of the light's waves. Fig. 2 shows the apparatus and its working.

Fig. 2

Fig. 3

Fig. 3 shows the relative ways of light's beams. If we lean the speech on Fig. 3 and we keep in mind that, following the experimenters' presuppositions, given that the run space of the beam from A to B to K was different (greater) than that from A to Q to K, when the beams would have been collected into the interferometer should have had to show an unfasement of the light's waves. But such unfasement didn't appear and then, as Feynman, we are compelled to accept the idea proposed by Einstein of a modifiable space as a consequence of the system's motion, and let follow the Feynman reasoning. He says that the interferometer didn't register any unfasement because the beam that made the distance A–B–K lasted the same time as that of the A–Q–K distance. Consequently, stated C the light's speed in the still system, if O–B (arms' length) is $OB = OQ = l$, and $A - O = OK = vt$, we shall have $AB = BK = l = Ct$, then $ABK = 2Ct = 2l$, while in longitudinal sense:

$$AQK = AO + OQ + QK = vt + l_1 + (l_1 - vt) = 2l_1$$

Then: $$\frac{AQK}{ABK} = \frac{2l_1}{2l} = \frac{\sqrt{l^2 - v^2 t^2}}{l} = \sqrt{\frac{c^2 t^2 - v^2 t^2}{c^2 t^2}} = \sqrt{1 - v^2/c^2} \quad (4)$$

And here we have the LF. So Feynman appears to have justified it as transformation factor of analog greatnesses of different systems relatively moving. But it is not so, because the found factor does not put in relation greatnesses of different systems but different greatnesses of the same system (in this case the one still). It put in relation, indeed the longitudinal spaces (in the direction of the motion) with the transversal ones (perpendicular to the motion's direction). But this tells us nothing. What we need are instead the relations between OB and AB, between AQ and OQ and between QK and OQ, that are analogue greatnesses of different systems. Then $OB/AB = l_1/e$. Then, as already seen, $l_1/e = \sqrt{1 - v^2/c^2}$, that however warths only for the greatnesses transverse at to the motion.

This conclusion answers to the fourth disagreement with the Einsteinian relativity (argument that will be treated better ahead in this book).

About the relations between AQ and OQ and QK and OQ, it would be a mistake to look for them confronting the complete spaces AQK and $2OQ$ because this does not make appear what is meaningful at this analysis point. It is that the going space of the beam towards Q, shows an enlargement of the still space with respect to that in motion and that takes place behind the moving space. The entity of this enlargement is given by $AQ/OA = (l_1 + vt_1)/l_1 = 1 + vt_1/Ct_1 = 1 + v/c$ which is the (2) relation already found. Analogically, for the beam's going back space, there is a contraction of the still space and that happens ahead of the moving space. Such contraction is given by $QK/l_1 = (l_1 - vt_1)/l_1 = e - vt_1/ct_1 = 1 - v/c$. This is the (1) relation already seen.

Disagreement No. 2

This last speech on space modifications ahead and behind the systems' origins answers partially the why of no. 2 disagreement with Einsteinian relativity. For a full answer, we must examine what happens to a moving body not identifiable with the origin point of a coordinate's system, but that has a given length.

Let's return to the simple case of two systems, one of which is moving as in Fig. 5

Fig. 5

Let's suppose too that on the moving system solidly joined to the origin O, there would be a body of a length, lying on the motion axis, with an extreme in O_1 and the other ahead in O_{11}. If, at first, we consider the moving system applied in O_1, a would be part of the compressed space and then would be compressed (shortened). But if we consider the moving system applied in O_{11}, a would be part of the enlarged space and then extended by the motion. So, to know the true modified length of a, it is necessary to decide where to apply the origin of the moving system (if in O_1 or O_{11}). But this origin is the point that comes to coincide with O when starts the light's beam on which we base our speeches. We have then, for a body in motion of a length, so the chance to result shortened as that to result extended, but if we take into account the particular case of O_1 and O_{11} coincident and applied in middle position between the extremes of a in O . . ., then the ahead part of a would shorten as $a_a = (\frac{1}{2})a(1 - v/c)$, whilst the back part would extend as $a_b = (\frac{1}{2})a(1 + v/c)$. The new dimension of a would be then $a_n = a_a + a_b = (\frac{1}{2})a[(1 - v/c) + 1 + v/c)]a_n = (\frac{1}{2})a[2]; a_n = a$.

This result makes us conclude that it is not true that the motion causes the bodies' contraction. Instead it causes the modification of the space around the moving bodies following the ways shown in Chapter 5.

Then in Chapter 5 we will have the complete answer to the No. 2 disagreement with Einstein's relativity.

Disagreement No. 3

It is wrong to sustain the inverse variation of space and time because this is contrary to the principle of light's speed constant. Such principle implies indeed the constant of the rate $ds/dt = C$, where ds and dt are differentials (variations) of space and time (space run by a light's beam in the time dt).

Disagreement No. 4

To sustain that space–time modifications happen only in the verse, and direction of bodies' motion derives, presumably, by having taken the *Morley–Michelson* experiment as general condition for C measurement. That experiment, instead works in a very particular condition. It measures the light's speed in only two directions: one of reference, perpendicular to the motion, and the other in the motion's direction. That such experiment should be done that way was perfectly justified because its purpose was to verify eventual differences of C which would have been highest when the measurement directions would have been following the motion and perpendicular to it.

But when verified the C equality in both directions, it is obtained the C constant principle. Then it must worth for every direction and verse, besides being independent from bodies speed. Moreover, the same Einstein, in the demonstration from which he obtains his four fundamental relations of restricted relativity (5), starts assuming the case of light running in whatever direction. It is at the demonstration's conclusion that the space–time transformation appears only in the motion's direction. Then it is necessary to examine Einstein's relations and their correctness.

Chapter 4

The Mathematical Way to Obtain the LF

We saw the Feynman way to obtain the LF. We could call it the physical way. We will now see one that we could call a mathematical one because it uses prevailing concepts of linear algebra.

This method is reported by two texts for physics students: one is *Elementi di fisica sperimentale-parte* 1ª, of L. Colli and U. Facchini, and the other is *Istituzioni di fisica teorica* of P. Caldirola (*viscontea* edit.), which makes the same treating too in the text: *Elettromagnetismo*, edited by *Tamburini E Masson*.

We will follow specially the text Colli–Facchini that, at the presentation of the relations obtained at the treating, states, 'These equations represent the famous Lorentz transformations in their most simple form and have been obtained the way described by Einstein and by himself put to fundament of relativistic mechanic.' See here the equations, $X = \dfrac{X_1 + vt}{\sqrt{1 - v^2/c^2}}$; $y = y_1; z = z_1;$ $t = \dfrac{t_1 + X_1 \dfrac{c}{c^2}}{\sqrt{1 - v^2/c^2}}$ where $t = x/c$. Indeed, being $t_1 = X_1/c$ we have: $(x_1 + vt_1)/c = t_1 + vx_1/c^2$.

Caldirola, like me, prefers the following system, equivalent to the (5). (It's enough resolve the (5) in function of x_1, y_1, z_1, t_1 to obtain utL

$$X_1 = \frac{X - vt}{\sqrt{1 - v^2/c^2}}; \quad y_1 < y; \quad z_1 < z; \quad t_1 = \frac{t - \frac{xv}{c^2}}{\sqrt{1 - v^2/c^2}} \qquad (6)$$

The reason of the preference of the (6) system is for the simplification of the following speech remaining in line with the other already done.

Speech already done is that that has given $X_1 = X - vt$ instead of the value that appears at the (6) system. Following the Colli–Facchini text the (5) and (6) equations are in their simplest form, then they must be adapted too to the very simple case by us considered in which $y = y_1 = 0$, $z = z_1 = 0$ and then $x = ct$. This situation is that represented in Fig. 1, of which we take into consideration the high right part. If we look for the value of X_1 starting from the (6) equations, we find, as already put in evidence $X_1 = (X - vt)/\text{LF}$, where X, in this case is $X = Ct$. The Lorentz–Einstein's relation, then, does not mirror the prospected situation. But we can hypothesise over possible situations to which it would adapt. The best one would result by considering our monodimensional situation as projection of a more complex situation representable in two dimensions as shown in Fig. 6

Fig. 6

Following the figure it is necessary to note two things. The first is the arbitrariness of the choice of the perpendicularity of the vectors vt, and $\overline{X_1}$; the second is that the transformation factor is not defined and we don't know if it is LF. We define it starting from the relation $\overline{X_1} = X_1/TF$, $TF = X_1/\overline{X_1}$; X_1 is known because it is $X_1 = X - vt$ and $\overline{X_1} = \sqrt{c^2 t^2 - v^2 t_1^2}$, given that $\overline{X_1}$, ct, vt_1 are sides of a rectangular triangle. Vt_1 is not known, but with it we can build the relation: $vt/vt_1 = vt_1/X$ from which: $(vt_1)^2 = X \cdot vt$. Substituting this value of $(vt_1)^2$ in the relation that gives $\overline{X_1}$, we will have: $\overline{X_1} = \sqrt{c^2 t^2 - Xvt}$ and being $X = Ct$, $\overline{X_1} = t\sqrt{c^2 - cv}$. The rate $X_1/\overline{X_1}$ will be then

$$X_1/\overline{X_1} = t(c-v)/t\sqrt{c^2 - cv} = \sqrt{(c-v)^2/c(c-v)} = \sqrt{(c-v)/c}$$

Then $TF = \sqrt{1 - v/c}$, which is *not* the Lorentz factor. Now it is evident that the application of LF in our case would be possible only at the cost of two arbitres and one mistake. The first arbitre has been of considering a phenomenon perfectly describable in one dimension, as needing a bidimensional description. The second arbitre has been of considering necessarily perpendicular the vt_1 and $\overline{X_1}$ vectors (in the executed representation; in the following the perpendicularity will be imposed to the vectors vt and $\overline{X_1}$).

In the following treating, the mistake will consist in considering the $\overline{vt}$ vector and its projected vt vector as being the same greatness, so when they define the projected greatness $X - vt$, as when vt serves to calculate the transformation factor.

Let's see then how we can obtain the LF in this case and the mistake we are compelled to make. Let's make reference to Fig. 7

Fig. 7

On it we have anew the representation on a plane of the three vectors ct, vt, $\overline{X_1}$ that correspond to the vectors Ct, $\overline{vt}$ and X_1 of the monodimensional representation. Following the Einstein relation (the first of the (6), which would give the value of X_1 (for us $\overline{X_1}$

)) the v appearing at numerator would be the same appearing at denominator under square root. But it cannot be so; indeed for that relation to give the value of X_1 starting from X (then C and t) and from V, this last must have at numerator a different value than at denominator in the LF.

This speech bases itself on the presupposition that the time t is equal in vt as in $\overline{vt}$, but this is an imposed condition by the fact that we are looking for values valid in O, starting from those valid in O. The time, then, must be the unique time t of O.

Let's then calculate the transformation factor on the basis of the over declared presupposed: $TF = X_1 / \overline{X}_1$; $X_1 = Ct - \overline{vt}$. Let's draw $\overline{vt}$ from the relation: $\overline{vt}/vt = vt/ct$ based on the similitude of the triangles: $O - O_1 - \overline{O}_1$ and $O - \overline{O}_1 - P$. In this relation we have just had to consider different the values of $\overline{vt}$ and vt, against the implicit assumption of the Einstein relation. Then $\overline{vt} = v^2 t^2 / ct$; $\overline{vt} = v^2 t / c$, then: $X_1 = ct - v^2 t/c$;

then: ???

Then:

???, which is the Lorentz factor, but obtained with an erroneous proceeding.

We have seen that Einstein's relations are not adapted to a monodimensional representation, but we can think this to depend by the fact to have assumed $y = y_1 = 0$ and $z = z_1 = 0$ and consequently $X = Ct_1$. We can think, it is, those equations to work in a more general case, as that of a bidimensional representation. Of this aim it must be said that the bidimensional representation can be considered, for our problems, the most general, because we have always three vectors to put in relation and always disposed to form a triangle. A triangle, being a plain figure, is representable in two dimensions.

The three overstated vectors are Ct = travelled distance of a light's beam started at $t = 0$ time from 0 system and arrived at the P point at the $t \neq 0$ time; vt = travelled distance of 0 system, started from 0 with v speed, accomplished with the same $t \neq 0$, in whatever direction from that of the light's beam; $ct_1 = c_1 t$ = travelled distance of light's beam from O_1 to P. The equivalence $Ct_1 = C_1 t$ is right because, for the determination of the travelled distance of the light in O_1 can be assumed or the t time in O and then light's speed must be considered altered in $C_1 \neq C$, either the constant of C and then must be considered altered the time from t to t_1.

The operation to do is, anyway, that to determine the Ct_1 vector knowing Ct, vt, and the angle between Ct and vt, or alternatively, to determine Ct knowing Ct_1, vt e the overstated angle (or that between vt and ct_1). When determining those vectors, the ratio Ct_1/Ct or its inverse is the TF for the space–time greatnesses from the still system to the moving one, or vice-versa. We can see that, laying the speech on Fig. 8, this

Fig. 8

way of calculating leads us to different results from those of relations (6).

By Einstein, indeed, (relations (6)), the motion of O_1 relatively to O, causes, and only in the X direction, a contraction, only ahead given by ???. But this reasoning is only half right, because the introduction of the LF, which we saw obtained from the Morley–Michelson experiment, postulates a space–time modification only in vt direction, when, in Fig. 8, X_1 is not seeable and should be looked for, without reason, in a three-dimensional representation.

By our Fig. 8, instead, we can choose between two TF – the one that make pass from ct to C_1 and that which make pass from X to $X - vt$. This last would be TF = $(x - vt)/x$ which is a banal result because it would give $X_1 = x - vt$. But this would not be only a banal operation, but too meaningless and wrong if we want the so obtained TF to give us the values, along the x axis, of the space–time modifications due to O_1 motion along the same axis x. These modifications are made seeable, instead, in Fig. 8, by the tensors Ct_1, it is by the vectors O_1P_3, O_1P_2, O_1P_1, for different directions from vt, and by O_1P_0 for the case we are treating. Then, drawing the TF in function of Ct, while for us it would change of value, for given x and vt, at the variation of ???, by the position we are criticising it would remain unvaried and only referred to the vt's direction. But also, in this case, being in general $x - vt \neq O_1P_0$, the TF would be different.

Here we have shown that it is wrong to think the $(X - vt)$ vector so related to X, as to ct, to give the space–time modifications in vt directions; it remains to show that that idea is meaningless and pernicious too, because it hides a phenomenon of precious consequences than the simple contraction of moving objects, either of the space ahead of the objects but only in the same direction.

Chapter 5

The General Transformation Factor

To give that demonstration, let's build Fig. 9, starting from Fig. 8. In it appears

Fig. 9

four systems: the O (still), the O_1, O_{11}, the O_{111} moving with respect to O in the direction given by x–x axis, and with speed given respectively by the vectors: O–O_1, O–O_{11}, O–O_{111}. The figure is the representation in one of the infinite planes passing through x–x of the sphere defined by all the OP vectors, going to all directions and of length $l = Ct$. The l_1, l_{11}, l_{111} vectors are, instead, the ones pertaining to the O_1, O_{11}, O_{111} systems corresponding to the l vectors. We will call 'tensors' such vectors because they give, with their length the contraction or extension space-temporal in all directions and depending on the relative speed of the systems to O system.

Fig. 9, then, says what happens to space–time as a consequence of the relative systems' motion. The value of space–time modifications may be obtained making reference to Fig. 8. As a consequence of what over said, we will define as TF the rate $Ct_1/Ct = C_1t/Ct$. Let's then start to draw C_1t from Fig. 8.

???

But, if we want the TF in function of $\beta = \pi - \delta$, we will have, from Carnot theorem:

???

Equations (8), (9), and (9 bis) are equivalent and we can verify that they give the general TF to obtain the values of spaces and times in a system in function of spaces and times of another system moving relatively with respect to it.

Let's make verifications for the following particular cases:
a) c and c_1 in the same direction of v and $v = -c$, $v = 0$, $v = +c$.
b) C_1 perpendicular to v and $v = -c$
   $V = 0$, $V = +c$
c) C_1 perpendicular to $V \neq -c$, $v \neq 0$, $v \neq +c$

a) case
   Relation (8)
   If C and C_1 are in the same direction of v, then, for $v = -c$, $\cos \alpha = -1$ while for $v = c$, $\cos \alpha = 1$ and for $V = 0$, $\cos \alpha = 0$.
   Then, for ???

   Relations (9) and (9 bis)
   If C and C_1 are in the same direction of v, then for $V = -C$, $\cos \beta = -\cos \alpha = 0$. (The same for $V = +c$), while for $V = 0$, $\cos \beta = -\cos \alpha = \pm 1$. Then, for ???

b) case
   Relation (8)
   If C_1 is perpendicular to V and ???

   Relations (9) and (9 bis)

If C_1 is perpendicular to V, we have always $\cos \beta = -\cos \alpha = 0$. Then for every value of V we have: ???

c) case
Relation (8)
If v is perpendicular to C_1, $\cos \alpha = v/c$, then for every value of α and v we have:
???

Relations (9) and (9 bis)
If V is perpendicular to C_1, then $\cos \beta = -\cos \alpha = 0$, so in any case: ???

So we have verified the relations (8), (9), and (9 bis) to be equivalent as they give the same results in different cases, but at parity of conditions. Moreover, they are consistent with restricted relativity geometry as represented at Figs. 8 and 9.

It remains to be shown that the (8) and (9) relations, in cases of c, v and c_1 laying on the same direction and $-c < v < +c$, give the already found relations (1), (2), and (3).

In those cases, the (8) has ???

The perfect equality with the relations (1), (2), and (3) is evident keeping in count that so c as v can have so positive as negative verse.

At this point it is necessary to underline the differences between our general transformation factor and that of Einstein, because these differences drive to important consequences not allowed by the Lorentz factor

Chapter 6

The Transformations From the Factors of Transformations

The first difference between our TF and the LF consists in the fact that the TF applies to tensors of a still space to obtain the corresponding tensors of the space modified by the motion of a body in still space; modifications that happen in every direction as shown by Fig. 9. The LF, on the contrary, applies only to the motion direction and gives the dimension modifications of the moving body only in that direction.

The second difference is that our TF attribute to the transformed tensors the negative sign too, whilst, by the same Einstein, the negative sign of the LF would have no meaning.

The third difference, at last, is that, applied to the speed, the LF gives birth to a relativistic dynamic, for which, the reaching of light's speed by a body would be impossible, because it should require an infinite energy. We will see ahead that this idea is unjustified.

We have already treated the first difference and seen its consequences, then it's unuseful to treat anew.

It's important instead to discuss the fact of the TF having a negative sign. To see to what drives such sign let's recall the (c) case of the preceding paragraph. This case has the particularity to put the transformed tensor perpendicular to motion's direction and then in intermediate direction of δ angle from O and π. This gives the chance to extend the conclusions about the negative sign to all the tensors oriented as δ. It must be premised that, as δ intervenes in the TF relation with its cosine, its sign is not important, because it is always $\cos \delta = \cos - \delta$ (see 9 bis relation).

Given the overstated premises, let's lay the speech on Fig. 10, where the C_1 tensors, transformed from the C, are positive if they lay on the superior part of the figure and negative if on the inferior one (arbitrarily). By the same arbitrary way let's establish as positive the V vector when at the right side of y–y axis, and negative in contrary case.

Fig. 10

Then we can see that if $V = 0$, for $\alpha = (½) \pi$ (Relation 8), with the increasing of V, C_1 diminishes as $\sin \alpha$, while C rotates as α at its decreasing from $\alpha = (½)\pi$ to $\alpha = 0$.

When $V = C$, $\alpha = 0$ and $C_1 = 0$, but we have no reason to think C not being able to rotate too towards negative values of α, which, as seeable in Fig. 10, give negative values of C_1. Negative values of C_1 bear to negative values of the time for the bodies moving along those tensors if we consider positive the run spaces. Indeed we have: ???. It must be noted anyway that the C_1 sign inversion affects only those C_1 tensors which arrive to annul for $V = C$. These are only the ones, that, in the case we are treating, have $\delta \leq (½)\pi$.

At this point we must discuss the rotation of C, which we assumed continuous in spite of the fact that it depends on V, which, at its turn, can't overtake the c value, but on the contrary, to allow the c rotation, a time reached the c value, must begin to diminish.

We can give a satisfactory explication of this phenomenon if we think the moving body (O_1 system) as revolving around a given point, while V and C_1 maintain their directions although changing verse.

Let's see how the things work in Fig. 11, where we have the body O_1 revolving around the P point. The body is represented in four perpendicular positions where V and C_1 reach maximum and minimum values in reciprocal opposition, as C components, in the constant direction x–x for v and y–y for C_1, while c rotates perpendicular to revolution's radius.

If we start to examine the motion by O_1 position at left, there we have $V = 0$ and $C_1 = C$. At the successive position (Pos. II) it is $V = C$ and $C_1 = O$. At the following position (III) it is anew: $V = 0$, but $C_1 = -C$. At last (Pos. IV) it is $V = +C$, while $C_1 = O$. All this corresponds to what we have seen to verify in the discussion about the negative sign of C_1. Indeed we see C_1 to become negative as consequence of V reaching C and to return positive when V becomes $+C$ (having C inverted the motion's sense). But V too, has here inverted the motion's sense (but not the direction).

For the phenomenon to develop as described, it should be necessary, however, to have the O_1 system rigidly vinculated to a P point, whilst it is not. And in reality, for O_1 to can describe a perfect circumference in the space has to be verified the particular condition that V, which is the only motion's parameter directly controllable, shall vary in the time following the law: $v/c = tv/2\pi R$ (10) where t is the time for the O_1 system to accomplish a total cycle revolving at light's speed along the R' radius circumference (Fig. 11).

But, if instead of a perfect circumference, we content of any cycle (complete when C retakes the original verse, not being required to return to the exact starting point) we can obtain it simply making run O_1 in straight line up when it reaches C speed, up to make it stop after, then, inverting the motion' sense, up to make it retake the C speed (now $-c$) and finally making it stop anew.

At this point, anyway, we must be acquainted of some things. At first Fig. 11 does not give the going over of O_1 system in the space. Indeed that is the diagram of the speeds along the cycle, which, eventually, must be seen as an half, circumference, run at first in a verse and after in the opposite one.

Secondly, too the half circumference form is not real because we have derived it from Fig. 10 without taking count of the fact that the rotation of C because of the variation of V must be seen, at the same time, in the two opposite verses (orary and antiorary). As a consequence, Fig. 11 diagram should be completed by a circumference symmetric to that drawn with respect to the K–K axis. But, this way, O_1 would result to have to take, at the same time, two diverging runs. We resolve the problem thinking to be the space to rotate sideway to O_1 system, while this moves straightaway in a cyclic run of going and returning. Hastily we note that we have here the explanation of the analogue quantistic phenomenon by which a fast moving particle can be found, at the same time, at two very far opposite points. In reality these two points are far only for the observer in the laboratory space, whilst they are coincident in the particle's motion space, which, for the C speed, results, sideway, symmetrically, contracted, according the ???

At last we must not forget that the speeches made are based on the simplification to have reduced phenomena which happen in a three-dimensional space to a bidimensional space. Then, in reality, the cyclical motion of O_1 system bears the cyclical rotation, around the run line of O_1 of a toroidal shaped space.

Chapter 7

Relativistic Dynamic

We have seen that the chance to enter into a negative space–time, which offers, it is, the chance to go back to past times, depends on the possibility to reach, by a moving system, the light's speed. But following the Einsteinian relativity, this would be impossible because of the nature of the LF, which applied to a body's speed, implies, for $V = C$, the expense of an infinite energy. Let's see the things precisely. The energy (kinetic) of a body of m mass in function of its speed V is: ???. According to Einstein, however, V must be transformed with the LF. Then: ???

From our viewpoint, instead, the things stay different. To understand it let's begin to note that, to apply the LF to the speed V as Einstein does, it means to apply it to the space run by the moving body, but not to the time. Applying it to the time too, V would not undergo transformation. Indeed: ???

Keeping this in count, we note that Fig. 9 shows an analogue situation, because the V, as the C, are drown as spaces run in the same t time valid in the O system. The same thing worths, obviously, for the C_1 speeds, which Being the passing to negative space–time a fundamental phenomena of the relativity here sustained, it is the case to deepen the speech showing what happens to space and energy too during the cycle we have examined for the speeds. For that purpose let us make reference to Fig. 10 and let's suppose the v speed to be determined by constant acceleration following the relations: $v = at$. Then the spaces run by O_1 body will be given by: $s = (½)at^2$ and the kinetic energy by: $E_c = (½)mv^2$; $E_c = (½)ma^2t^2$.

Then if we normalise the relations putting $a = m = c = 1$, we have that the v vector of Fig. 10 represents the times valid in O system, that we can assume to be too the ones of a detached observer.

Then if we represent in a Fig. 10 bis the values of s and E_c, at the varying of v, putting them perpendicular to the v values, on y–y axis, we have a sinusoid that tells us several things.

The first thing is that the sinusoid itself can be made only operating two reversements of the part of the figure corresponding to where $O \leq v \leq c$ and the C_1 are negative. This means that the speed had to be considered in diminution, because of the reversing of the x axis, after $v = c$, but not the run spaces, notwithstanding the fact that, how it will be better seen at Chapter 7, the spaces must be considered, at that phase, as negative, after the reaching of C by the moving body. In that phase, v must be considered diminishing as the kinetic energy, that, on the contrary in Fig. 10 bis results max1 for $v = 0$ in negative space–time. But this, anyway, helps to understand that the represented energy may be considered as correct if instead of the simple kinetic energy we think it as the total necessary energy to make the body reach the C speed and then brake it up to $v = 0$.

The second thing the drawing tell us is that it is a mutilated representation because it takes into account only the half plane of the positive x whilst the real situation must collect all the planes (complete ones) crossing the x–x axis.

A third thing to note is that if the body is not braked, just by inertia law, reached c, it would persist on its motion for ever. This means that if the body would be a watch, for infinitesimal times by it measured, infinite times would run onto the its ahead tensor.

A fourth thing has to do with the space around the body. This, together with the time, undergoes contractions and enlargements expressed by the C_1, related to the C, as at Fig. 10, when the C_1 is perpendicular to v. The corresponding Fdt is: ???

This speech, which confirms previous interpretations of relativistic phenomena, explains the fact that photons could run enormous tracks and continuously oscillates between the O and C speeds. The reality is that the oscillation is only apparent (where there are no gravitational fields to accelerate or slow down the motion) because the above considerations tell us the photon constituting PAEs have enough energy for the completion of an only cycle. Then it is necessary to think that they collect energy (E_c) at the emission, that they conserve it along the track (where they then don't oscillate) and that, at last they give it away at track end when they are stopped. The phenomena of interception of light as beam of a given frequency must be then explained as result of the emission of the PAE at those frequencies, fact which find explanation as quantum phenomena, given it is known that energy emission, at subatomic levels, is made precisely in an uncontinous way and with quanta of value determined by emitting body temperature.

??? have different lengths just because of having been traced on the base of the time valid in O and not in O_1, O_{11}, O_{111}. Fig. 9, then, if given the chance to obtain the kinetic energy of a body moving at V speed, the drawn value is the right one to substitute that of the (11).

But Fig. 9 is made just on the base of not transformed C and V speeds, because are speed in a considered still space and not modified by any system's motion in it.

The speed to take into account for the calculation of a moving body's kinetic energy in a still space is V in the relation $E = (½)mV^2$ (12), where m = mass of the body and V its speed measured in the still space. When $V = C$, $E = (½)mC^2$, which is not an infinite value. Then = *it is theoretically possible, for a body, to reach the light's speed.*

As a corollary to this conclusion, we can ask ourselves what might happen to a body to which should be impressed a greater energy than the one necessary to reach the light's speed. Such energy's surplus might not be translated into a speed's increment because C is a limit speed. Then? Einstein's answer was this: *The energy's surplus goes to increment the body's mass.* Our answer is this: *The energy's surplus brakes the body.* But this, after the body reaches the light's speed. But how does this happen? Because, as we saw, reaching C, the body enters (as to say) into negative space–time. The negative space–time is characterised by negative spaces and times so that the following relation are realised: $ds^-/dt^- = V^+$; $ds^-/(dt^-)2 = a^-$ (ds^-, dt^- = differentials of space and time; V^+ and a^- = positive speed and negative acceleration). It happens then that the body passing to the negative space–time conserves the positive verse of the speed, whilst it inverts the acceleration, braking its own motion.

If we think the braking will bring the body to the $V = 0$, the braking energy must worth $E_b = (½)mC^2$ for the max V equal to C. But E_b equals the kinetic energy of the body moving with $V = C$, then at the stop the body will have received and dissipated the energy $E_t = 2(½)mC^2$; $E_t = mC^2$ (13) that corresponds to the completing of half cycle, because at the stop the body is again in negative space–time. It is to complete the cycle returning to $V = 0$ in positive space–time that the body needs a energy's surplus equal to that already spent during the first half cycle. Altogether these energies shall worth $E_c = 4(½)mC^2 = 2mC^2$ (13 bis)

This relation is analog to the famous one of Einstein, but it does not say the same thing, because that gave the body's intrinsic energy, with no respect to its motion, whilst this gives the needed energy to accomplish a total cycle in space–time, starting and arriving with zero speed and reaching two times the C speed.

Moreover, this speech shows that an energy's surplus over $E = (½)mC^2$, brakes the body.

Moreover, the acquisition of negative space–time idea orients to understand, as we will see more ahead, that furnishing to a body a greater energy then ($½mC^2$, does not conduct to mass increment, but to the beginning of an oscillating motion of the body between C and V (inferior to C) speeds.

Here we can add that the E_t energy, which annuls at the cycle's end, is transferred to the space in a form that explains the photon's nature. It is important to repeat that without the negative space–time possibility, the photon's oscillating nature does not find explanation. We will see this better, more ahead.

Chapter 8

Mass and Speed

It is not enough to have shown that a body's speed in a still space does not undergo a relativistic transformation and then, that it is possible it reaches the light's speed with the finished energy's spending $E = (½)mC^2$.

It is not enough because Einstein tells us that the mass m is not constant, varying with the speed V by the relation $m = m_0/LF$ (14) which, with the (13) draws to $E = m_0C^2/LF$, where LF, if $V = C$, is LF = 0. Then $E = \infty$. But how did Einstein obtain the (14) relation? See here his proceeding. He gives the relativistic definition of a m_0 momentum $q = m_0V/LF$.

At this point, it must be noted that it is plausible to apply the LF to the V, because it is the transformation factor found to regulate relativistic relations among spaces, times, and then speeds. Then, with a simple mathematical passage, q becomes $q = m_0V(1 - V^2/C^2)^{-½}$. Developing in series, the binomial $(1 - V^2/C^2)^{-½}$ becomes $\left(1 + \frac{1}{2}\frac{V^2}{C^2} + \frac{3}{8}\frac{V^4}{C^4} + L\right)$. Then Einstein divided arbitrarily q by V and obtained: $m = m_0\left(1 + \frac{1}{2}\frac{V^2}{C^2} + \frac{3}{8}\frac{V^4}{C^4} + L\right)$ where the LT is no more applied to the speed but to the mass without physical reason. At last, in a way completely arbitrary, m is multiplied by C^2 and gives the result: $mC^2 = m_0C^2 + \frac{1}{2}m_0V^2 + \frac{3}{8}m_0\frac{V^4}{C^4} + L$. Stopping the series at the second term we have at last: $mC^2 = m_0C^2 + \frac{1}{2}m_0V^2 = m_0C^2/LF$. At this point, Einstein interpreted the m_0C^2 term as value of the m_0 mass intrinsic energy and $(\frac{1}{2})m_0V^2$ as the kinetic energy due to V speed. But on which base?

It is possible to show that m_0C^2 is the m_0 mass intrinsic energy if it is supposed to be rotating around its own mass center with peripheral speed $V_t = C$. Such rotational energy is indeed: $E_r = E_t + E_p$ (E_t = kinetic energy due to C tangential speed; E_p = potential energy of the mass detained by its own attraction force). But $E_t = E_p$ and $E_t = (\frac{1}{2})m_0C^2$, then: $E_r = 2(\frac{1}{2})m_0C^2 = m_0C^2$.

It has been possible to show the rightness of that relation, but not following the Einstein's proceeding whose logic escapes us.

About the $(\frac{1}{2})mV^2$ term, we all can interpret it as the classical kinetic energy of m_0 mass at V speed, but our interpretation does not prove the proceeding's validity from which came the term,

Moreover, if we can put $MC^2 = M_0C^2/LF$ and to have so: $MC^2 \to \infty$ for $V \to C$, not the same we can do with $m_0C^2 + (\frac{1}{2})m_0V^2$, which, for $V \to C$ cannot result equal to m_0C^2/LF. Indeed, for $V \to C$, $m_0C^2 + (\frac{1}{2})m_0V^2 = 1.5m_0C^2$, whilst $M_0C^2/LF = \infty$.

Anyway, in spite of the arbitrary and not understandable proceeding that drove to the (14) relation and in spite of the now seen inexactitudes, the (14) boasts important confirmations so at electrically charged particles acceleration processes, as at experimental verifications.

To show the incorrectness of (14), we must then explain why bringing mass to light's speed does not require an infinite energy.

It is necessary, then, to analyse at first the charged particles' acceleration process, to understand how the (14) accounts for it, even if it is possible to exclude the increment of mass as the impossibility to reach the speed C. On the contrary, it is the possibility to reach C for the particles, and then their entering in negative space–time that makes necessary the energy's surplus that the (15) expresses. Let's see accurately the things referring us to Fig. 12, obtained from Fig. 10.

Fig. 12

Fig. 12 sketch justifies too the relations derived from the (16) one and particularly the (19/b), because it shows that for $dv \to 0$ and ds proportional to dv, $dt = ds/v \to 0$ too, as there assumed, when ds and dv are space's and velocity's differentials in correspondence of $V \to C$.

From Fig. 12 we obtain that the passing from V to C speed implies the passing from ???.

The needed energy for the particles increment of speed from V to C is: $dE = (½)mC^2 - (½)mV^2 = (½)m(C^2 - V^2)$; $dE = (½)mC^2$. This says that for $V = 0$ and then $C??? = C$, $dE = (½)mC^2$, while for $V = C$, $dE = 0$. This would make think to facility to bring the particle to the light's speed, but it is not so because we have: ???

This relation, for $V = C$ gives: $C???/dV = \infty$

Now, as $C???$ is a function of dV, consequently, to obtain $C??? = 0$, it should be necessary to regulate V with absolute precision, which is practically impossible. Then V is normally smaller than C, so when $E < (½)mC^2$, as when it is $E > (½)mC^2$, because in such case, as already seen, the energy's surplus produces the mass deceleration. Trying, as it should be done, to reduce the dV differential, it is necessary to increase the force that pushes the particle, it is the accelerator's power. This operation causes a consequent reduction of C???, at every cycle up to when lasts the power increment. At the same time, we have reduction of the cycle time's duration. This proceeding could theoretically go on forever, bringing the average particle's speed (that will oscillate between V and C) ever nearer to C, but then the accelerator power should reach infinite values. Indeed we have:

$$W = f.V = maV = m(c_1/dt)V \qquad (17)$$

where W = acceleration power
f = pushing particle force
a = particle's acceleration
dt = time for e_1 to reach
value = time for V to reach C

Then we have: $dt = t\,dV/C$ where t = time for the particle to pass from O to C with the average acceleration: $a = C/t$. Then:

???

and for the (16): ???
which for $V = 0$ gives $W = 0$
and for $V = C$ gives $W = \infty$.

But in the case we would calculate the power (theoretical) to bring the particle to C speed, avoiding the passage to negative space–time and then the oscillation, we have only to substitute dt with t, for the (17) becomes ???; $W = mCV/t$ (19), because, in this case $C_1 = C$. After we have $C/t = a$ then: $W = maV$ (20) which, for $V = 0$ gives $W = 0$ and for $V = C$ gives $W = maC$ which, for m and a finite values gives W as a finite value.

If now we draw the power from the Einsteinian energy's relation (15), we have:

???, which, for
$c/t = a$ becomes:

??? (21) that for $V = 0$, gives $W = m_0 ac$ while, for $V = C$ gives $W \cong \infty$. If now we confront the (18) with the (21) in the case of $V = C$, we constate the equivalence of the two relations, but while the (21) gives no explanation for the infinite value of the power (the null value of the LF at denominator we have already seen to have no explanation), the (18) tells that such infinite value of the power appears only if one does not take into account that when $V = C$, $dt \rightarrow 0$.

Then, if on this base, we calculate the energy that the maximum power produces during the dt time, we have (from the (19)): $E = dt\,W$, which for $W = \infty$ gives: $E = O \times \infty$ (19/b) which is not necessarily an infinite value (see Fig. 12 and following note).

At this point, we have shown our relativistic dynamic gives count of particles' decelerator functioning at least so well as the Einsteinian one, but we can ameliorate our valuations by examining the experiments of M. M. Rogers, Reynolds, and F. T. Rogers (from now RRR), driven with an apparatus purposely conceived to measure the dependence of mass on speed.

Speaks us of this experiment Piero Caldirola in his 'istituzioni di fisica teorica' explaining us that the R.R.R. apparatus based on the same principle exploited by J. J. Thomson, but it gave more accurate results and such to can say that Einsteinian relation that tights mass to speed is verified inside the 1% of experimental values. Very better then, than the Abraham relation:
???
where $\beta = V/C$. Relation, this one, which had, historically, great importance, but came out defeated from the R.R.R. experiment

because gave values that overtook of about ten times the experimental error.

Now if we take for good the Caldirola's information, we must take for granted too the diagram ($\beta - m/m_0$) that the book shows to prove the statements. But at this point we are in embarrassment because the diagram bears three experimental points the value of which must be measured on the page because they are not given by the text, while we are in a condition to judge the correctness of the Einsteinian curb, that on the diagram results altered (ad hoc to not diverge from 1% of tolerated mistakes), whilst, corrected, at a point, it diverges of 4.79% relatively to the experimental value. Moreover, it must be noted that the experiment takes into account a fraction very narrow and of small meaning of the field of particles speed ($0.62 < \beta < 0.758$); it is only a 13.8% and moreover where the mistake is greatest is at the point of $\beta = 0.758$, showing a clear tendency to divergence from calculated curb and experimental values, so to give the certitude that, for V values more near to C, the divergence could even overtake 1000%, which leads to the disqualification of Abraham relation.

Anyway the experiment would show the dependence of mass on speed and this does not copy with our views. Then it is necessary to examine deeply the experiment and its presupposed beginning by the experiment of J. J. Thomason.

Chapter 9

Thomson's Experiment

Thomson's experiment was not thought just to verify the dependence of mass on speed but to measure the ratio e/m between charge and mass of electrically charged particles (electrons and protons, mainly). Only after it was seen the apparatus could work well also to verify the mass by speed dependence. We can see why following Carlo Castagnoli text 'Lezioni di struitura della materia'. The apparatus is shown schematically at Fig. 13.

All the apparatus is in extreme void and is constituted of a wire K for electrons emission whose energy is regulated by means of A lattice. D is a striver to aim the electrons to the O point of the fluorescent screen when electrical ??? and ??? magnetic fields, both regulable, are null. The fluorescent screen is disposed in a way to measure X and Y coordinates of the point where the electronic beam hits the same screen at the varying of electrons' speed and of ??? and ??? field's intensity. X and Y coordinates are in relation with ???, ???, V (electrons' speed), e (electron's charge) and d, as follow:

$$X = eEd^2 / 2mV^2 \qquad (22)$$
$$Y = eBd^2 / 2mCV \qquad (23)$$

These two relations give the chance to obtain the ratio e/m because, for E and B maintained constant at given values, there are a couple of values X and Y which draw parabolas on the screen at the only vary of V. This V, unknown in principle, is obtained by the ratio Y/X that from (22) and (23) gives: $Y/X = BV/CE$. Then:

$$V = YCE / XB \qquad (24)$$

Obtaining V this way it is possible to obtain m/e by both the (22) and (23) relations, but this does not concern us. Instead, it is interesting for us to note that X and Y coordinates give the electrons energy so function of V as of m (at the same time). The electron's classical kinetic energy is $H = (½)mV^2$, while the energy furnished to the electrons by the apparatus results grater. Then it is as m or V should be greater than the values for H determination. But the apparatus does not distinguish between an oscillating V and one that is not oscillating. Supposing V is not oscillating, the energy's excess is automatically attributed to a hypothetical increment of their mass. But if V is oscillating, the energy's increment has a natural explanation just in the energy's expense required by the oscillation, just as seen when analysing the need of power for the working of charged particles accelerators. The Thomson experiment, then, does not clear the problem of mass–speed dependence more than do the particles' accelerators.

Let's see then, if the RRR experiment can help us.

Chapter 10

The Rogers, Reynolds, Rogers Experiment

The apparatus used for this experiment differs from that of Thomson because the deviation of the electrons in the ??? field is obtained by means of a radial condensator instead of one with parallel armatures, which makes possible a more accurate reading of the X value. Moreover, the electrons emission is not created by a thermoionic process, that does not permit a very accurate control of furnished energy, but is obtained by β radiation of $Ra(B+C)$, which consists of three different and well-defined energies.

Although nothing changes substantially with respect to Thomson experiment, because if now we have more accurate values of X, Y, V and H, we don't know yet if the exceeding energy relatively to classic relation is due to mass increment or to speed oscillation, nevertheless this experiment gives the chance to reach an interesting conclusion if we put in diagram power and speed of particles following the different relations in discussion and experimental data. The powers interesting to us are those given by (18), (20), and (21) which we normalise putting to unity the m_0, a and c, while for the speed we take the ratio V/C. The experimental data are given as ratios m/m_0 in function of V/C. The ratio m/m_0 is obtained by the relation (21) as ratio from the power with $V > 0$ and $V = 0$ and is then: $m/m_0 =$

1/???. The experimental data are then directly confrontable with the curb of the power of (21) relation, which when normalised becomes $W = 1???$. The (21) curb, at last, results directly confrontable with the (18) and (20) ones, because are all curbs of powers of the same particles, normalised the same way for the same parameters. The spoken diagram is the following:

Fig. 14

Before discussing the interesting conclusion we can draw from Fig. 14, we must analyse the consequences of the fundamental difference of functioning between the Thomson apparatus and that of RRR. Such difference consists in the fact that the electrons which go to hit the screen in the first case are accelerated along all the *d* run (Fig. 13), whilst in the second one don't undergo acceleration after the emission. They are emitted by the source at the maximum energy which we can think they can't lose during the run to the screen (they move in extreme void). Then, on Fig. 14 diagram, they must appear only as three points (one for each radiation beam, which power is its energy referred to an unitary time). Then those power values will be expressed by the same energy value obtained from RRR experiment. These values are put on the diagram and the standard interpretation sees them, experimental mistakes excepted, on Einstein curb (21 relation). Here, instead, is proposed an interpretation which keeps in count the fact that a shooted by an atom electron has an energy resulting by its revolution around the nucleus. This is $E_t = E_s + E_p$, where $E_s = (\frac{1}{2})mV_2$ is the kinetic energy due to the revolution speed V, while E_p = potential energy = E_s.

$E_t = 2E_v = mV^2$.

If we confront this energy (power) with that of the relation (20) of the diagram, we see it worths the double at every point with equal V. But the experimental RRR data, result with double energy (power) with respect to the (20), with only little differences that, at once, make us to side with our interpretation; moreover, if we keep in count the experimental data of which we dispose are not original, but obtained by measuring on a text diagram. The following table shows the differences from the confrontations.

Table

The table shows the powers calculated with the (21) relation give more than double differences with respect to observed powers than those calculated with the (20) relation we support. This drive us to refuse the Einsteinian relation $m = m_0$??? on which the (21) is found and which would a mass of a moving body to increase with the speed. Then the necessary energy's increment to increase the particles speed in the accelerators is not explained by the mass increment but by the fact that the accelerated particles, when they reach the light speed, begin to oscillate between speed C and an inferior one, with growing frequency and proportional growing energy's absorption, as we will see ahead.

It remains to show that too the fact of (21) relation to give as starting power for a mass acceleration a $W > 0$ proofs the speed relativistic oscillation and that samely proof the experimental data of R.R.R. To do this it is necessary, at first, to study the relation between kinetic energy and speed of a moving mass when relativistic causes intervene to render oscillating the speed.

Chapter 11

Oscillating Speed Energy

Fig. 15

Let's lay the speech on Fig. 15 on which mass m starting from 0 towards X^+ reaches a stable speed C after three oscillations that bring it successively to the speeds $V_1, V_2, V_3 = C$, crossing C every time and going to reach, with the braking fases, the $V_{(0-1)}, V_{(1-2)}, V_{(2-3)}$ speeds. The V values are determined as if for each half oscillation the angular run of it should be reduced of d$\propto$.

If now we generalise the case of Fig. 15 putting any n instead of 12 and we look for the relation that gives the energy of furnish to the mass to reach stably C, we have:

$$E_t = \sum E_u \qquad (25)$$

where E_t = total energy
E_u = energy of each single
oscillation from V_i to C to
$V_{(i+1)}$ to C to $V_{(i+2)}$;
to each unity of i corresponds d$\propto$.

$E_u = 2[(½)mC^2 - (½)mV_i^2 + (½)mC^2 - (½)mV^2_{(i+1)}]$;
$E_u = [mC^2 - (½)m(V_i^2 + V^2_{(i+1)})]$;
$V_i^2 = C^2 \sin^2(n_i d\propto) = C^2(1 - \cos^2(n_i d\propto))$;
$V^2_{(i+1)} = C^2 \sin^2(n_{(i+1)} d\propto)$;

$V^2_{(i+1)} = C^2[1 - \cos^2(n_{(i+1)}d\alpha)];$
$E_u = 2\{mC^2 - (\tfrac{1}{2})mC^2[\sin^2(n_i d\alpha) + \sin^2(n_{(i+1)}d\alpha)]\}$
$E_u = 2\{mC^2 - (\tfrac{1}{2})mC^2 \times 2[1 - \cos^2(2n_i d\alpha)]\}$
$E_u = 2\{mC^2 - [1 - \sin^2(2n_i d\alpha)]\};$
$E_u = 2(mC^2 - [1 - (1 - \cos^2(2n_i d\alpha))]\};$
$E_u = 2[mC^2 \cos^2(2n_i d\alpha)].$

Retaking the (25) we have:
$E_t = \sum E_u = 2mC^2 \sum \cos^2(2n_i d\alpha);$
$\sum \cos^2(2n_i d\alpha) = [()\sin(2\alpha) + (\tfrac{1}{2})\alpha]/d\alpha$ and for $d\alpha = \pi/n$ and $V \to C$:
$\sum \cos^2 \alpha = 2\{[2n((\tfrac{1}{4})\sin \pi + (\tfrac{1}{2})(\pi/2))]/\pi\};$
$\sum \cos^2 \alpha = 2\{[2n(0 + (\tfrac{1}{4})\pi)]/\pi\};$
$2 \sum \cos^2 \alpha = n.$ Then:
$\sum E_u = nmC^2$ and for the (25)
$E_t = nmC^2$ (26)

We can verify the relation for Fig. 15 case, where $\alpha = 180°$, $d\alpha = 15°$, $n = 12$. There we have $\sum \cos^2 \alpha = (0.5 + 0.9330127 + 0.75 + 0.5 + 0.25 + 0.0669873) \times 2 = 6$ and then $2\sum \cos^2 \alpha = 12 = n$, while $n = \pi/d\alpha = 180°/15° = 12$. The most interesting characteristic of this relation is that E_t is not in function of V as in the case of the Einsteinian relation (15), but of the discreet value of n, which brings us to draw some important conclusions which make us proceed so to the demonstration of our relativity validity, as to make light on Known, but not enough cleared, physical phenomena.

At first it must be noted that in the case of $n = 1$ ($V = 0$), E_t is not 0, but $E_t = 1mC^2$, but now the reason is clear because E_t is just the energy to bring the mass from $V = 0$ to $V = C'_0$, when $C'_0 = C$. In the (15) relation instead, the initial energy should be the mass intrinsic energy; it is not understandable why it must be furnished if it is already possessed.

The other conclusion to which one arrives is that the energy to furnish to m for $V \to C$, differently from what the (15) says, is not infinite, but a function of n, which can be also 0. In such a case, being 0 the number of the masse cycles around C, $E_t = 0$, as the (26) says. Up to here then, our relativity confirms the classical physics because it shows the chance to reach the light's speed with finite energy, but we can see that it throws light too on quantum physics because it confirms that the energy is transmitted by discrete quantities (discrete, indeed are the n values).

The commission between quantum physics and relativistic dynamics can be pushed more deeply when m is thought equal to a PAE such that the (26) might be written:

$$E_{ve} = hv = n \text{ PAE } C^2 \qquad (27)$$

where E_{ve} = photon energy
h = Planck's constant
v = photon frequency.

h is an action constant. It is: $h = E_u \, dt$ (energy of one oscillation cycle by a time unity); $v = n/dt$ then: $hv = E_u \, dt \, n/dt = E_u n$; $E_u n = n$ PAE $C^2 = E_{ve}$.

In the case of Einsteinian relativistic dynamic: $m = n$ PAE and $E_{ve} = mC^2$

The photon, then, is a small pack of nPAE moving with $V \to C$.

To have proposed this view worths to Einstein, in 1921, the Nobel award, which was not bestowed him, instead, for what he had previously formulated of greater importance by his view point: the special relativity (at the 1905) and the general one (1915).

Chapter 12

Fossil Radiation Uniformity

The fossil radiation which arrives to us, with great uniformity, from every part of the universe (as the experiments Boomerang and COBE have proved) arrives too from each point at the same temperature of 2.7 °K. Now, if the uniformity of the radiation from everywhere can be explained by the cosmic expansion, so much that Gomow, foresaw it before to be discovered by case from Wilson and Penzias (technicians of the Bell Telephones) during an antenna installation, the temperature uniformity represents a problem. This because, being the universe expansion, the bodies farther from each other go off at a proportionally greater speed. Then, on Earth, radiations that started from the farthest body must arrive hottest because the radiation photons, moving in an almost perfect void, don't undergo any energy attenuation. The attenuation of the radiation now arriving at 2.7 °K is, then, due only to the cooling of the emitting material. To this regard is thought the cosmic material to have began to irradiate when the cosmic broth reached (towards the low) the 3,000 °K temperature, such, it is, to permit the atoms formation. If we assume this view, the farther cosmic material, as said over, should send the hotter radiation, because it was emitted when the material was hotter; instead there is no temperature difference from reaching us photons, independently by run length. Why? Because, given the

Earth (it is, what represented it before its formation) of Big Bang instant, was inside the exploding cosmic broth, if we idealise it with a P_0 point, each other around point, found itself to go off from it, by the explosion effect, with a speed proportional to the distance from it, in the very logical supposition that the explosion energy would have uniformerly distributed in all the broth. Let's take then ten point plus one, which at hydrogen formation (around 300,000 years after the Big Bang) should be at equal distance the one by the other and going off from the Earth (point (P_0) in a straight line, and with a speed of 1/10 of light's between each point and its near one. Let us also suppose that such speed has remained constant for ten thousand million years following the Big Bang. Let us give marks from 0 to 10 to the said points starting from the Earth. If now we suppose to be at ten thousand millions years from the beginning of the cosmic radiation, and we direct a telescopic antenna to collect the radiations coming from those points, the ones ten thousand millions years cold are the ones emitted at P_0 point (then from near Earth cosmic material), whilst all other points of our alignment should send us hotter radiations, following the done speech, if the relativity would not intervene to change things. Indeed, following our relativity, a body going off from another modifies space and time, behind itself, depending on the $TF = C^1/C = (C+V)/C$, where V is the body's going off speed. Applying the TF to the time we can write: $T_t = T_0/TF + t/TF$ (T_t = total time from the beginning of the cooling of emitting mass to today; T_0 = time from the beginning of the cooling of emitting mass to the instant of the emission of P_i).

T_i = time from P_i emission to the instant of reaching the telescope.
Then: $T_i = T_0(V/C)$;
$T_t = (T_0/TF)(1 + V/C)$;
$T_t = T_0$

This equality says that the radiations arriving at the same time to the telescope (P_0) come from all the P_i points which at the instant of the first atom formation had the same temperature, but moreover they arrive as if they would be emitted after the emitting materials same cooling time and then all at the same temperature. This important result could not be obtained with the Einsteinian relativity because its transformation factor is the LF which would give a time T_t: $T_t = (T_0/???)(1 + V/C)$ which would be, always $T_t \neq T_0$. Making an example with numbers can help better to understand the speech.

Let P_3 be a body going off from P_0 at the speed $V_3 = 0.3C$.

It causes, behind itself a space–time modification given by TF = $C'/C = 1 + 0,3C/C = 1,3$

If P_0 mass underwent a cooling time of ten thousand millions years, the P_3 mass would undergo a cooling time $T_3 = 10/1,3$; $T_3 = 7.6323$ thousand millions years. Then the radiation, to run the three thousand million light years to reach the P_0 point, takes the time $T_p = 3/1,3 = 2,3077$ thousand millions years, which added to T_3 gives: $T_t = T_3 + T_p$; $T_t = 7,6923 + 2,3077 = 10,000$ millions years, which is the time for the material cooling at the P_3 point, lets us arrive it radiation starting from the instant at which its temperature reached the 3,000 °K so allowing the atoms formation. This radiation then arrives together with that emitted by P_0 but does not arrive hotter as if emitted by a three-thousand-million-years younger material because the 7,6923 thousand millions years is the time passed in P_3 whilst, in P_0 which is in the space–time modified by P_3 motion, the time results: $T_0 = 7,6923 \times 1,3$; $T_0 = 10,000$ millions years, which is just equal to T_t.

The reasoning made for the P_3 point, works logically for all the points around the Earth, which, at every distance irradiate the energy they had when they were at 3,000 °K of temperature. Then all the fossil radiation that arrives and will arrive at the same time to the Earth, now and forever, will be at the same temperature. But if this worths for the fossil radiation does not work the light radiation arriving from stars which shows a 'red shift' so greater as farther its origin and then faster its going off speed. Why? Because of the difference of the stars, which emit light of same frequency (if and when so) not depending on the distance of observation, the fossil radiation material emits temperatures so greater as farther it is, perfectly neutralising (as shown) the shift towards greater wave lengths of the radiation.

Chapter 13

Fizeau–Zeemann Experiment

This experiment was already executed in 1851 to obtain proofs of ether existence. Indeed, Poincaré said that the experiment did touch with hand the ether existence. When Einstein knew it, he prayed God to forgive him. He, indeed, had an interpretation of the experiment results based on speed relativistic composition and that excluded the ether necessity. But both were wrong because the experiment correct interpretation nor shows ether existence neither owes anything to Einsteinian relativity. Let us see the things closely. That experiment was driven on an apparatus as that of the sketch of Fig. 16.

Fig. 16

That apparatus is constituted of two parallel pipes of same length on which water is made to run as shown, with w speed in the pipes. A monochromatic light beam (b) is sent on an half transparent mirror S which splits the beam and sends the two resulting beams to Σ_1 and Σ_2 mirrors, which reflect them to the pipes. Coming out from the pipes a total reflection prism (or mirrors) send anew the beams to the pipes, but now in reversed verse. At the coming out, the mirrors Σ_1, Σ_2 and finally the two beams to I interferometer for the examination of interference fringes. It must be kept in mind that the two beams run paths of equal length, then with still water, without ether eventually drawn by water, the beams did not show fringes at the interferometer. They appeared, instead, with water in motion. But they did not correspond to what was expected in a situation characterised by Galilean relativity and would have been given the total drawing of either. In as such case, indeed, had to result the following in water path times: $t_1 = 2l/(v - w)$ and $t_2 = 2l/(v + w)$, where l = pipes length and V = light speed in water (still one).

$V = c/\eta$ where η = light refraction index in water with respect to air or void. $\eta = 4/3$. The difference of time employed by the two beams in the water paths had then to be: $\tau = t_1 - t_2$; $\tau = 2l$??? but, as said, the interference fringes gave a different result that was interpreted as w was modified by the motion according to the relation: $w^1 = w(1 - 1/\eta^2)$.

Why was this interpretation given? Because it was fit to be explained with the speed composition according to the Einstein's special relativity. But now we can see that that interpretation was wrong. Indeed, it puts the light speed in water as:

$$V = [(c/\eta) \pm w]/[1 \pm (\beta/\eta)] \tag{30}$$

where $\beta = w/c$, but this can not be, because being $V = c/\eta$, the (30) becomes:

$c/ = [(c \pm \eta w)/\eta]/[\eta \pm (w/c)]/\eta$;

$c/ = [(c \pm \eta w)]/\eta]/[(c\eta \pm w)/c\eta]$;

$c/ = (c^2 \pm cw)/(c\eta \pm w)$. Now multiplying both members of the relation by η/c we have: $1 = (c\eta \pm \eta^2 w)/(c\eta \pm w)$, which can be only in case of $\eta = 1$. The mistake can be only at the (30) relation. Indeed it comes out from having assumed by definition: $dx/dt = vx$ and having put, after, $vx = v$, when, putting us in relativistic perspective (constant c), it must be $dx/dt = c$. Indeed, with $\eta = 1$ we have, starting from (30),

$V = (c \pm w)/[1 \pm (w/c)]$; $V = (c \pm w)/[(c \pm w)/c] = C(c \pm w)/(c \pm w)$. As a conclusion, $V = C$, which must not be.

Moreover, the (30) relation owes nothing to Einsteinian relativity because results by having put V as ratio between a $dx = (dx^1 + wdt^1)/???$ And a $dt = [dt^1 + (\beta/c)dx^1]/???$. But this ratio makes the roots ??? disappear which are the (Lorentz) factor of transformation of special Einsteinian relativity for space–time greatnesses.

Anyway it seems that the V, defined by the (30), satisfies the requirements of the relation that gives the τ. It is necessary, then, to find out a relation to substitute the (30), giving the same values for the τ. We can do it starting from the following (30) transformation: $V = c/\eta \pm w (1 - 1/\eta^2)$. Putting numbers to the relation we have: $V = 225,000 \pm w (0,4375)$. The new relation to be found will have to substitute the $(1 - 1/\eta^2)$ bracket with another, giving the same result (0,4375) and, obviously, being physically justified.

We must, anyway, remember that the $(1 - 1/\eta^2)$ bracket had been accepted because drove to a theoretical result enough near the experimental one, and not just for some kind of physical–mathematical evidence.

The following arguments will bring to a little different result of the bracket, but enough near to that to substitute to can think to be it the right one. Unluckily Fizeau and Zeemann have not transmitted us the experimental value corresponding to their bracket (or, at list, the books did not do it).

The following considerations start from the presupposition that, if the light in the water runs with a different speed than in the void it's because we are in the presence of a relativistic phenomenon by which the spaces where the light moves, result contracted by $1/\eta$ with respect to the void. This can be explained by the fact that, being in the water, a great motion of particles in every direction and at great speed occurs. The $1/\eta$ contraction in every direction results into a three-dimensional contraction of the space equal to $1/\eta^3$. A so contracted space, having to fill a non-contracted one (that of the pipes, in our case) and having to do it at expenses of any only dimension, this shall result contracted just of $1/\eta^3$.

Again in our case, $1/\eta^3$ being the contraction of the water column, addition of that of the two pipes (21) is too $1/\eta^3$ the relativistic contraction to which undergoes the water speed w in the pipes. This, then, becomes: $w^1 = w(1/\eta^3) = w(0,421875)$.

The 0,421875 value, then, if the made speech is correct should substitute the 0,4375 one so as the $(1/\eta^3)$ bracket should substitute the $(1 - 1/\eta^2)$. Let's commit to good will experimenters the honour and burden to verify, while we may subject to experiment too with different liquid from water or with solid transparent materials as glass or plastic bars which permit to eliminate the distortions due to the turbulent motion of the water, as the difficulties to measure its speed.

Chapter 14

Black Holes Evaporation

If the black holes evaporate, as Hawking says, the thing has for us a special meaning because we are able to explain such phenomenon, not yet actually explained. The black holes are not really holes, but very concentrated masses and are black because they don't emit any radiation. Then they are invisible. Only recently proofs have been detected of their existence as lenticular effects; it is as deviation of light beams coming from behind them. These deviations show a circular deformed space around an invisible point which is supposed to consist in an enormous and enormously dense mass just having the black hole characters. The main characteristic of black holes is its enormous attraction force with which they catch every body passing near, but too the photons, which being massless should escape from their attraction. This is the legend, which, however has no justification in classical physics, and then it is not true the phenomenon must be explained by the general relativity as it is often said. It is true that the photons, being massless don't give rise to any attraction force as says the Newton's universal gravitation law: $f = MmG/r2$ (31)

where f = attraction force

M = Big attracting mass

m = attracted mass

G = gravitational constant

r = distance between M and m

Then if $m = 0$ too $f = 0$, but another Newton's law states that $f = ma$ \hfill (32)

and then $a = f/m$. In the case as ours, even if f and m are 0, a can be $a \neq 0$. Moreover, in practical cases, a value can be easily obtained from not zero values of f and m. Knowing a of a given gravitational field, the classical cynematic tells that by that acceleration effect all the bodies attracted in that field get a speed given by the Galileo relation $v = at$, independently by their mass. The photons too, then, in a gravitational field moves as the other bodies, but at the difference of them, given their own nature, if emitted or reflected by the body generating the gravitational field render it visible. The fact of the black holes to be invisible means then that they neither emit nor reflect photons, and this because of their very great gravitational acceleration. Let's go to see the things more near considering that the gravitational acceleration due to mass M, following the (31) and (32) is $a_g = MG/d^2$, while the centrifugal acceleration of a body stably revolving around it is: $a_c = V^2/d$, where V is the peripheral revolution speed and d is the revolution radius. If we suppose the photon to revolve very near the M spherical surface, we can write, $V^2/r = MG/r^2$ where $r = d$. If M is spherical, as supposed, its value will be: $M = (4/3)\pi r^3 \delta$, where δ is the density. If, then, we suppose V to be $= C$ (case of the photon), the (33) becomes: $C^2 = (4/3)\pi r^2 \delta G$, where the only unknown greatnesses are r and δ. At the end we have: $r^2\delta = 3.2261 \times 10^{27}$ g/cm, and this is the condition over which an M increment, translating into an $r^2\delta$ increment, cause the photon catching with its absorption. But under this value we have still black holes, it is, conditions in which the photons, although not absorbed, are not emitted to space because their speed is lower than the escape one;

they revolve around M with different values of $r_r > r$ (r_r = revolution radius when greater than the r of M).

The limit value of $r^2\delta$ for the photons to escape from M mass is obtained, keeping in count that the escape speed for a body revolving around M in equilibrium between attraction and centrifugal force and with revolution radius = r of M (supposed spherical) is $V_f = \sqrt{2}V_r$ (V_f = escape) (speed; V_r = revolution speed). As $V_f = C$, the value to put into the (33) at the place of C will be $V_r = C/\sqrt{2}$. The (33) then becomes $C^2/2r = MG/r^2$, from which we can obtain the new value of $r^2\delta = 1,6130 \times 10^{27}$ g/cm. This is the minimum value for a spherical mass of δ density and r radius can be a black hole.

In case such density would be of about 6 g/cm³, similar to the Earth's one, the black hole radius should be 4 = 1,7961 × 10¹³ cm = 179,613,855 km; it is equal to 1.2014 times the distance Earth–Sun.

But we know that big masses can collapse reaching densities of the order of 10^{15} g/cm³. In such case, the minimum black hole should have the radius $r = \sqrt{1,613056 \times 10^{12}}$ cm = 12,7006 km.

Up to this point the speech has had only the purpose to show the black holes reality can be supposed also starting by classical physics, it is, without recur to general relativity. But now we can show our relativity can explain the Hawking thesis by which the black holes, in spite of $r^2\delta$ values greater than 3.226×10^{27} g/cm, evaporate. To speak the truth, we must say they evaporate just because they can reach such values of $r^2\delta$, because in such conditions the engulfed matter reaches the light speed, passing to negative space–time, where it accomplishes the braking and speed inversion operation. This is then the inverted process of catching, which happens however in negative space–time and then does not appear at the present where we observers are.

This inverted process consists in sending back to the origin the material and radiation arrived. In this meaning, the black holes actuate the conservation of information.

But our speech doesn't work at 100% because the black holes don't swallow only photons but whole stars too, and those must be visible during their run towards the hole, because even if they reach the light speed, must be verified the principle of the light speed constant. In other words, if those stars emit photons towards the Earth, those photons, even if the stars should run at speed C farthing from Earth, would arrive us at speed C, making visible the stars. Which, all them, running towards the hole from every direction, should hid it as very brilliant heart of a galaxy. This phenomenon is known and observable.

To explain the existence of invisible black holes we must, then, introduce a new hypothesis, that too founded on our relativity. This hypothesis puts the holes into a far future when the stars they will swallow shall be already extinguished because of the conclusion of their life cycle.

Black holes of this type will be the remnants of dead galaxies. They will not evaporate because their constituent material will have never reached the light speed to pass to negative space–time. Preconstituted reason for this different behaviour will be the fact of these black holes having formed at the center of small galaxies which stars would have nor the space and neither the time to reach the C before to be swallowed.

The problems about the future exploration will be treated ahead in Chapter 19.

Chapter 15

Matter Formation

We face the matter formation problem so for its importance for the propulsion of space–time transportation means, as because gives us the way to verify, a time more, the validity of our relativity theory. For this purpose let's make the well-founded hypothesis that the subatomic particles are constituted of AEP (Absolutely Elementary Particles, See *note* 1 of the *introduction*). We found this hypothesis on the famous relativistic–quantistic relation:

$$hv = m_0 C^2 \qquad (51)$$

where $h = 6.625 \times 10^{-27}$ erg·s is the Plank's constant, while v is the photon frequency, the energy (intrinsic) of which is given by the Einsteinian relation $E = m_0 C^2$. The hv product, in the case of $v = 1$ gives the energy of 1 cycle of oscillation of m_0 mass. This mass, then, will be obtained by (51) as follows: $m_0 = h/C^2$, which in CGS units gives:

$$m_0 = 6{,}625 \times 10^{-27}/(g \times 10^{20}) = 7{,}361 \times 10^{-48} \text{ g}.$$

If we confront this mass with that of the electron, $m_e = 8{,}99 \times 10^{-28}$ g, we have the ratio: 1.22128×10^{20} which is the number of AEP constituting the electron.

If we go on with the idea to see the matter constituted of AEP, let's suppose the aggregation force to be the most general, it is the gravitational one. On the same line let's suppose to obtain the electron dimension by the relation of its intrinsic energy: $E_e = m_e C^2$, valid in the case of the AEP constituting the electron are revolving around the axis of the same with light speed. The electron form shall be then that of a rigid, round necklace, whose forces acting on the AEP shall be $f_c = mC^2/r_{ee}$ (centrifugal force tending to spray the AEP to the space) and $f_a = mMG/r_e^2$ (attraction force on the AEP when at a distance r_e from the axis. In the relations, m = AEP mass; M = electron mass; r_{ee} = real electron radius; r_e = maximum electron radius. The difference between r_e and r_{ee} is that, while r_e determines the electron volume, being the distance from the mass center to the surfaces, with which shall be calculated the attraction force, r_{ee} is the value of effective radius on which depends the centrifugal force.

As the AEP are supposed to revolve around the electron mass center with elliptical orbits with half greater axis equal to r_e, we will have $r_{ee} = Qr_e$, with $Q < 1$. We will then write $f_e = f_a$ (valid for formed electron); $f_c = mC^2/Qr_e$; then $mC^2/Qr_e = mMG/r_e^2$; $C^2/Qr_e = MG/r_e^2$; $C^2 r_e/Q = MG$; $r_e/Q = MG/C^2$; $r_e/Q = $???; $r_e > 6{,}6526 \times 10^{-56}$ cm.

This is the radius the electron should have to exist when it would be constituted of AEP of a 7.361×10^{-48} g mass in the quantity of 1.22×10^{20}, revolving at light speed around the electron mass center.

Now, too supposing to be $Q = 1$, we have, at max, $r_e = 6{,}6526 \times 10^{-56}$ cm, while we know the electron real radius is of greatness order of 10^{-14} cm, very greater of the calculated value. How to explain this difference? By considering that, following our relativity, the space represented by the tensors perpendicular to the direction of a moving body undergo a contraction given by $C_1/C = $???. When $v \to CC_1/C \to 0$. In our case the value of the radius of the electron tends to zero which is exactly on the line of the tensors contracted by the AEP motion.

We can then conclude with the possibility that the electron radius to be the greatness order of 10^{-56} cm for relativistic reasons. This possibility increases if we take into account that the calculated electron attraction results linearly directly proportional to r_e radius, so that the ratio between the measured radius of 10^{-14} cm and the calculated one must give the ratio between the attraction forces (electrostatic/gravitational) measured by the Thomson experiment. That ratio was $k = 4.17 \times 10^{42}$. The ratio of our radiuses is $10^{-14}/10^{-56} = 10^{42}$.

This way, recurring to our relativity, we have found some conditions of electron existence, but not all them, and indeed, with the presupposed taken in count, the electron can't form, and this because the $f_e = f_a$ forces' equality is an instable balance's condition so that under the calculated r_e the centrifugal force overtakes the gravitational one forbidding the AEP aggregation, while, over the r_e radius, the AEP would not meet limits to their aggregation and would come to form electrons of tendentially infinite dimension. We could understand better the speech making reference to Fig. 17.

Fig. 17

The figure is the diagram putting in relation centrifugal and gravitational forces with the r_e electron radius (virtual). The curbs cross when $r_e = 6,6523 \times 10^{-56}$ cm, being f_e and f_a equal as intensity, but opposed as vers. So, for values of r_e smaller than that of balance of the forces, the centrifugal force is greater than the gravitational one, forbidding the electron formation, while the contrary would happen if the r_e could reach the $6,65 \times 10^{-56}$ cm value but it can't.

So we have found the right electron to explain its mysterious very great electrostatic force, but it can't exist.

To solve the problem helps us anew our relativity, which allowing the possibility for moving bodies to reach the light speed, shows their possibility to pass to negative space–time. Following the hypothesis made from the beginning of this chapter, the electrons should be made of AEP revolving at light speed. If only we think a very small energy surplus would have brought them over C, the electrons would exist in negative space time.

In this situation, besides space and time, invert, as consequence, the accelerations and then the forces. With the inversion of these ones, the electron gets repulsive electrostatic force, whilst the centrifugal force becomes attractive on the AEP, instead to spray them around. So being the things, if we look Fig. 17 diagram, we see that the attractive force (now the centripetal one) is greater than the repulsive (now gravitational one) for values of $r_e \to 0$, whilst for r_e values $> 6,6526 \times 10^{-56}$ cm, the repulsive force prevails over the attractive one, forbidding the electron dimension increment.

The most convincing proof of this speech is the electrostatic repulsive force of the electrons.

At this point, we could continue the speech on matter formation, showing how (repulsive) electrons arrive to form protons, how protons and electrons join to form neutrons, how this joins the protons to form atomic nuclei, and, at last, how electrons, protons, and neutrons together join to form atoms and molecules, always exploiting, being the case, the behaviour foreseeing the inversion of the forces at the crossing from a space time to the opposite one, which is characteristic of the relativity we support. However, it seams to be the case to rest to examine the atomic nuclei structure that comes out from the made hypothesis, because it could teach us the most rational way to arrive to the matter fusion control, without which the space–time vehicles construction would be impossible. For this purpose, we must accept the idea that the protons, in the nuclei, notwithstanding the fact of being made as necklaces of positrons, at their time made as AEP necklaces, revolving as seen before and strongly mutually attractive, do not mutually destroy because they are kept separated by the neutrons. This explains the why of the at least quantitative equality of protons with neutron (with the only exception of the hydrogen, because this element nucleus has only one proton).

So being the things, eliminating from the neutron, the electron that makes it neutre, it loses its isolating property and becomes itself a proton reacting with the other nucleus protons and developing the fusion atomic energy exploitable as thrust by means of a rocket engineering.

The energy furnished by each proton at the reaction will be: $E_p = m_p C^2 = 1,6525 \times 10^{-24} \times g \times 10^{20}$ erg. $E_p = 1,48725 \times 10^{-3}$ erg. Then, having to be implicated at least two protons the smallest fusion energy obtainable will be $E_{mf} = 2E_p = 2,9745 \times 10^{-3}$ erg. It is a very concentrated energy, given that it is produced by only $3,305 \times 10^{-24}$ g of matter. This gives 9×10^{20} erg/g or $9,17431 \times 10^{12}$ kg·m/g.

Here we can see the main problems for fusion controlling are more of quality than quantity order. It is necessary to reach the capacity to intervene on very limited matter quantities, hardly localisable because in very fast motion, and in very short times. Nevertheless, it is said that it can never be said never. One day, which we hope is not far, the means to win the challenge will be realised.

Chapter 16

Light Refraction

It might seem strange to speak of the light refraction as a phenomenon not completely cleared, but it is so! Let's see why, making reference to Fig. 18.

Fig. 18

Following Snell and Descartes' laws that describe the phenomenon, a monochromatic light beam which cuts in a transparent material surface, continues its run into the material, changing the propagation direction of an angle resulting from the constant ratio: $\eta = \sin i/\sin p$, where i and p are the angles made by the beam with the perpendicular to the separation plane of the materials, drawn from the beam intersection point with the same plane. Another law establishes that the perpendicular to separation plane lies on a same plane with both strokes of the fractured beam.

Now we know η to be the ratio between the light speed in the two materials, but this is not represented in Fig. 18, which is drawn following the Snell–Descartes laws. Indeed the figure shows equal vectors for cutting in beam as for the refracted one. If this gives a correct ratio between the light speed components parallel to the materials separation plane, not the same it gives for the perpendicular to the said plane and this is not logic. A logical representation would be indeed that of Fig. 19, where the cutting in beam and the refracted one are expressed by vectors of η ratio between but lying on the same straight line. But this condition doesn't show refraction. But the fracture of the light beam exists, and it has not yet had explanation. But we can find a satisfactory explanation looking to a representation showing what happens to the light speed when it moves in spaces modified by bodies' motions. This representation is that of Fig. 20, which shows a body O, moving with speed V along the X axis with positive vers, while emitting a light beam with β (positive) angle deviation with respect to X axis. Let P be the P point where the system sees the beam after a unitary time. Then the O–P tensor will be deviated by an α angle with respect to X axis.

If the ratio $OP/O_1P = \eta$, then too $\sin \beta \sin \alpha = \eta$. Moreover $\beta = i$ and $\alpha = p$.

But we have the explanation of the total reflection phenomenon too which appears when the OP tensor, by increment of α, brings $\beta = 90^0$ which is the limit value for V, OP and O_1P for a triangle. And $90°$ is too the limit value of i to have refraction. With the increment of i over $90°$, it is with the crossing of the separation plane by the cutting in beam, the refracted beam disappears and at its place a reflected beam, symmetrical with the cutting in one with respect to the perpendicular to separating plane.

Our space–time scheme, then, copies-explains perfectly the light refraction phenomenon up to the lateral consequence of the total reflection. Anyway there persists a perplexity due to the fact of our scheme to show that to greater light speed ($C > C_1$) corresponds smaller value of the relative angle sine. We have indeed $C/C_1 = \sin \beta / \sin \alpha$ and not $C/C_1 = \sin \alpha / \sin \beta$. But this last equation is not required for the demonstration. Instead what is required is the constant of the ratios; it is that to the constant of the ratio C/C_1 corresponds the constant of the sin α/sin β ratio. But this requirement is satisfied even if $C/C^1 = \eta$ and sin α/sin $\beta = 1/\eta$. Indeed to the constant of η corresponds the constant of $1/\eta$.

Chapter 17

How to Travel in Space–Time

To travel in space–time implies essentially to move in space because the motion creates time unphasements between the moving bodies and the still places. This happens too when the motion speed would be very low, even if, in such case, the time unphasement would be irrelevant to the purpose of time travel. Moreover, to go back in time it is necessary to reach the light speed. But there are other things to take into account. They are as follows:

1) The space modifications are a consequence of the fact that the space gets tended or compressed depending on being ahead or behind or sideway of the moving body.
2) The time modifications happen in the modified spaces following the same rules. Then if a space contracts to its half size the same happens to the time flowing there.
3) The modifications duration is illimited not stopping at the stopping of the body which determined them. The motion stopping makes end only the extending of them. Then the space–time unphasements between space–time points result everlasting to annul such unphasements it is necessary to operate reversely of the way employed to obtain them.
4) The space–time modifications do not affect the bodies causing them.

Chapter 18
Space–Time Travel Projecting

The project of a space–time travel supposes to can dispose of the adapted vehicle to transport, for relatively long times so human beings as robots to the desired space–time point. This vehicle must reach the light speed if the intention is the access to the negative space–time to go back in time. But a near speed light is too good for fast travels in positive space–time towards the future.

Even if this memory faces primarily the theoretical possibility of space–time travels, towards the end it also will confront the more practical problems relative to the vehicles, just to discover that too the vehicle's construction is of very less difficulty than it could be supposed taking only in count the enormous speed to reach.

At this point of the speech, anyway, we give the vehicle for disposable and we go to concentrate the attention on the pure and simple travel.

As for every travel we must know where we must arrive, given as discounted to know from where we start. But if this is true for Earth travels or near Earth ones, for space–time travels the starting point must be precisely established so specially as temporarily, because being in motion with respect to the arriving point, the run to accomplish varies before the start. The arrival point, instead, can be considered fixed in a given point of the universe.

Up to here, it has been spoken of to the future travels and to the past as presenting same problems. This is only partially true because of the difference with the to the past travels; the 'to the future' ones don't require to reach the light speed. Let's then start to study these.

Chapter 19

Travelling Towards the Future

To speak of travels towards the future make suppose to be our planet future, populated of human being, able, then, to modify, perfecting it, the mean rounding them and then able to offer to people coming from the present (the past of future people) information and means to solve problems already solved by them, but not by the present living people. But the possibilities offered by 'towards future' travels are many and not only to the presents utility. Has not to be excluded neither the importance to go to discover the fast future of the universe, of our galaxy, of the solar system and our planet, giving answer to the questions that troubled the minds of nineteenth century scientists, which reflected on the problem of our solar system stability. Robots for the future travels will give these kinds of answers in a relatively short time.

Anyway, to draw the most interesting conclusion for us, it is enough to project a short travel to the future ahead of one thousand years of our planet.

The time reference respect to which we will speak of future, or of past either of present, rests then set on our planet before the starting of the vehicle can modify the Earth space–time. Never forget, indeed, that a vehicle leaving the Earth enlarges space–time behind and contracts it ahead. But, as it has been noted that the space–time modifications don't affect the bodies causing them, we can too make reference to the vehicle time after its watches would be synchronised, before the start, with the Earth time.

Important paradoxical reflection to do is that a 'to future' travel implies reaching the space–time point where the Earth should arrive after a given time (1,000 years, in our case), in a shorter time, otherwise we could not speak of travel to future. Moreover, the arrival to the said point has to be on the Earth itself which, instead, is supposed to arrive in 1,000 years. These 1,000 years, then, is the time which would be measured on Earth if the vehicle motion would not modify the space–time between the extreme points of the travel (Earth at the start, Earth at the arrival). Instead the vehicle moves between the two points compressing the space–time towards the arrival point and enlarging it towards the starting point. Compression and enlargement is that of measure unities, so that ahead of the vehicle many kilometers can be compressed in a few metres (following the unaltered measures of the vehicle) and the same, proportionally happens to the time. The contrary happens behind the vehicle so that corresponding to each year flowing on the vehicle, half year flows on earth. To be more accurate must be said that the ratio one to half for the vehicle-Earth time flowing is right only when the vehicle reaches exactly the light speed. In this case the space–time compression ahead of the vehicle reaches an infinite value. To reach then a space–time point of finite value must be scrupulously avoided the light speed. But probably, for quantistic reasons the moving bodies are protected against this eventuality and when their kinetic energy would overcome that necessary to reach C, they would pass immediately to a speed lesser than C in negative space–time.

Let's forget for now these observations and retake the study of the problem to arrive 1,000 years ahead in the future. Let's calculate, at first, where would be the Earth at that moment in the future.

It goes towards Apice (Ercole costellation) with the speed $V = 19,500$ m/s. In 1,000 years it will run $S = 19,500 \times 100 \times 365,25 \times 24 \times 3600$ m; $S = 6,1537 \times 10^{14}$ m $= 6,1537 \times 10"$ km. If this space could be run at the speed C, it would take the time $t = 6,1537 \times 10"/300,000$ s; $t = 2,051,244$ s $= 23,74$ days $= 0,065$ years.

But for our vehicle, however charged the possibility to travel at the speed C for all the distance is to exclude. That speed, rather, should be reached gradually with constant acceleration which, for human charge, should be near to the Earth one. It should be reversed, then, to reach the zero speed at the arrival.

The time for this double operation will be $t_a = 2c/a$; $t_a = 60,00,000$ s, which is abundantly greater than the over calculated t ($t_a/t = 29,25$) ($t_a = 694,4$ days) for $a = 10$ m/s². But the relativity helps us, compelling us to keep in count the time contraction due to vehicle motion. Because of this we can write $t_a = 2\sum t_p$ where t_p are the elementary times modified by the motion, given by the relation $t_p = dt(1 - V/C)$ (where $V = |a|\eta i\, dt$ and dt could be chosen of 1"). If the total time not modified for acceleration and deceleration is $t_a = 2V/|a|$, the number of the tp that enters into calculations will be $n_a = t_a/dt$, to consider double with respect to that of the single phases of acceleration and deceleration of actuated with the same $|a|$.

Putting then $\eta = n_a/2$ we can write, $t_a = 2???\, Dt\, (1 - ???)$; $t_a = 2(n - \sum???)$ having put $dt = 1$ to have t_a in seconds. Then: $t_a = 2(\eta - ??? \sum \eta_i)$. Being: $\sum \eta_i = ???\, t_a = 2\, (\eta - ???)$ and lastly:

$$t_a = 2n\, (1 - ???) \tag{200}$$

Following the (200) if $V = C$ and $|a| = 10$, we have: $t_a = 29,999,999$ s it is $t_a = 347^a 5^h 20^1$ to which must be added the time for uniform motion that, not modified, is $t_v = T - ta$, where the is the time towards the future to be covered with the travel. If, as in our case, $T = 1,000$ years $= 3{,}15576 \times 10^{10}$ s; $t_v = 3{,}14976 \times 10^{10}$ s. We obtain the modified t_v by the relation $t_{vm} = t_v(1 - V/C)$ which for $V = C$ gives $t_{vm} = 0$. We can see, then, that for the future travels, the needed time doesn't depend on the time distance to cover, when would be possible travel at C speed, but only by the acceleration to reach C, which determines t_a by mean of the (200)

But we can see that t_a too can be put to zero, so in the case of $n = 0$ (banal condition of missing motion), as in the case of ??? $=1$; it is $a(n + 1) = 2c$, but taking in count that $n = n_a/2 = t_a/2 = v/a$, we have $a(n + 1) = a(v/a + 1)$, then $v + a = 2c$, which can give only $v = c$ and at same time $a = c$, conditions that could be neared only by infinitesimal mass particles. For our purposes, then, the acceleration and deceleration times t_a can't be annulled. They can, however, be greatly reduced playing on a and V values. It is soon visible the importance, then, to travel with great a values, but apparently with low V values. But this only apparently because V enters to determine t_a above all throw the t_p, which by their side depend by the TF $= (1 - v/c)$. So we see that for $V = C$ the t_p annul with the annulling of the TF.

To clear the done speech come useful two verifications to confront with the done calculation for $T = 1,000$ years, $a = 10$ m/s² and $V \to C$, which gave $t_t = t_a + t_{vm}$ (total time); $t_t = 29,999,999 + 0$ s; $t_t = 348^d 5^h 20^1$.

Let's do the first verification with always $T = 1,000$ and $V \to C$, but $a = 10,000$ m/s², which would be an acceleration bearable by robots. In this case, $t_a = 2c/10,000$ and $n = n_a/2$; $n = t_a/2$. Then $n = c/10,000$; $n = 30,000$ s, and $t_a = 60,000(1 - ???) = 29,999$ s; $t^a =$

8^h 20^1 (transformed acceleration time), Now being $V = C$, t_{vm} will be $= 0$ and $t_t = t_a + t_{vm} = 8^h$ 20^1 which is a time notably smaller than the confrontation time of 347^d 5^h 20^1. The second verification we do it maintaining $a = 10$ m/s² and putting instead $V = (½)C$. We have: $t_a = 300,000,000/10 = 30,000,000$ and $n = 15,000,000$. Then: $t_a = 30,000,000$ $(1 - ???)$; $t_a = 22,499,999.5$ s which is smaller than the t_a of the confrontation calculation. But, in this case t_{vm} is not zero, but $t_{vm} = t_v \times TF$; $t_{vm} = (T - t_a) \times TF$; $t_{vm} = (3,155 \times 10^{10} - 30,000,000) \times (1 - 0,5C/C)$; $t_{vm} = 1.57638 \times 10^{10}$; $t_t = t_a + t_{vm} = 22,499,999.5 + 1,57638 \times 10^{10}$ $t_t = 1,57863 \times 10^{10}$ s $= 500^y$ 86^d 19^h 20^1 which is a time notably greater than the t_t of the confrontation calculation.

As these two verifications rend clear and the upper scheme of Fig. 21 shows, to save time in the travels to the future must be chosen the highest possible accelerations and decelerations to reach the light speed and leave it to brake up to $V = 0$. Instead the speed C maintained by inertia between acceleration and deceleration brings to zero the remaining travel time.

Fig. 21

Now let's take into consideration the 'back in time' travels.

Chapter 20

Back in Time Travels

Travelling to go back in time must take into account the fact that, at the difference of travelling to the future, it is needed to pass to negative space–time. This can be obtained, as we already saw, by reaching, with a small energy excess, the light speed. This acceleration phase presents no differences with that of travels to the future except, just the need to reach C with that energy excess (which could be too of an only quantum, then negligible in the calculations, provided it would be guaranteed).

At this point it fits to simplify the (200), to adopt it to vehicle time t_v, to bear in mind that the space–time modifications don't affect the vehicle itself. This simplification drives to the elimination of the relativistic modification of the elementary dt times and then consists in putting $t_p = dt$ (instead of $t_p = dt(1 - v/c)$, which drives to a: $t_a = 2v/|a|$ (201).

Then for $v = c$ and $|a| = 10$ is $t_a = t_{av} = 2C/10$ and $t_{av} = 60{,}000{,}000$ s $= 694d^-_4$.

But now having to get the anterior tensor time for acceleration and braking, we must undouble the (200) to keep in count that, differently from the 'to the future' travels, the breaking, in negative space time, causes the acceleration sign change. We will have then $t_{am} = n(1 - ???)$ and $t_{fm} = n(1 - ???)$.

Given the very high n value, we can put $(n + 1) = n$. After, being $n = c/a$, we arrive to:

$$t_{am} = (c/a)(1 - \tfrac{1}{2}) \tag{202}$$
$$\text{and } t_{fm} = (c/-a)(1 + \tfrac{1}{2}) \tag{203}$$

If $|a| = 10$ m/s², then: $c/a = \pm 30{,}000{,}000$ s, $t_{am} = 0.5 \times 30{,}000{,}000 = 15{,}000{,}000$ s and $t_{fm} = 1.5 \times -30{,}000{,}000 = -45{,}000{,}000$ s

The total time will be then:

$T_{ta} = t_{am} + t_{fm} = -30{,}000{,}000'' = -347^d\ 5^h\ 20^1$.

If at the end of acceleration and breaking operations the vehicle would find the Earth, this would be set back to 30,000,000 s relative to the moment of the vehicle starting.

Too different is the elapsed time on the quitted Earth, which remains on the vehicle back tensor. We can find this time again by undoubling the (200) to bear in mind so that, this time, the acceleration changes because of the breaking, but because we have the $t_p = dt(1 + v/c)$.

Then for acceleration and breaking we have: $t_{am} = (c/a)(1 + \tfrac{1}{2})$ (positive acceleration) and $t_{fm} = (c/-a)(1 + \tfrac{1}{2})$ (negative acceleration).

Then, on the quit Earth the elapsed time for acceleration and breaking would be: $T_{a+f} = t_{am} + t_{fm}$ $T_{a+f} = 45{,}000{,}000'' - 45{,}000{,}000 = 0''$.

It must not be surprising that at the same instant, judged (measured) of 60,000,000" from the starting by the vehicle, on Earth would have passed or 30,000,000" backward either 0". Indeed the relativity, basing on the mysterious light speed constant, destroys the idea of contemporaryty.

If we want to calculate the corresponding times for vehicles with $|a| = 10{,}000$ m/s² (for robots or apparatuses), we must only substitute in the relations (201), (202), and (203) the new $|a|$ value. We will have then time to reach the Earth in negative space–time: $t_{ta} = t_{am} + t_{fm} =$ (½) 30,000 + (1,5) 30,000 = −30'000" = −8ʰ20'

Elapsed time on the vehicle is t_{av} = 60'000" = 16"40'

Elapsed time on quit Earth = 0"

But times and spaces run by inertia at speed C are infinite for infinitesimal vehicle times and spaces.

Let's see in Fig. 22 vehicle times for 'to past' travels.

Fig. 22

Chapter 21

Conclusions

Travelling in time must consist of three distinct phases: 1ª) Acceleration: to bring the vehicle speed as near as possible to light speed so to reduce at the maximum the travel times.

As long as the vehicle speed remains lower than the light's speed, the elapsing times will be positive, when they will be negative passing to the opposite space–time.

The relation giving the acceleration times is $n = 2c/a - a$ s, with the sign of n depending from that of a and then from that of space–time.

2ª) Displacement, whose times are given by the relation $t_s = (T - 2n)(1 - v/c)$. When $v = C$, we would have $t_s = 0$ for every T and n value and this is the main reason to travel at light speed.

3ª) Braking: necessary to reach the destination space–time point with the speed it has respectively to the vehicle starting point. Supposing the two points are relatively still, the braking time will be equal to the acceleration one, but it shall be anyway given by the same relation seen at the first phase with sign depending by space–time sign.

At this point, anyway, we know only how to go back or forward in time, but we don't know if the instant we will reach at the end of the travel will be conjugated to the space point occupied by the universe-point (mostly Earth) we need reach with the time point. Indeed we start with the presupposition of reaching the Earth at an established date and not to reach that date finding nothing at the arrival. This issue would, indeed, be compelled if on the vehicle nothing should be done to regulate its way. We can understand better the speech if we go to calculate the space distance of our vehicle after its advancement in time of t_a = 30,000,000 s. The relation that gives the space for constant acceleration motion is $s = \frac{1}{2}at^2$ if a = acceleration. In our case, with accelerated motion for half the time and decelerated for the other half the proper relation must be $s = 2[(\frac{1}{2})a(t_a/2)^2]$; $s = (\frac{1}{4})a \times 30,000,000^2$. If a = 10 m/s², s = 2,25 × 10^{15} m = 2,25 × 10^{12} km. The Earth distance, corresponding to s will be given, instead, by its speed towards Apice for the not modified time t = 1,000 years. It is at the speed of 19,500 m/s:

s_T = 1,000 × 365,25 × 24 × 3600 × 19,500 m

s_T = 6,153732 × 10^{44} m = 6,153732 × 10^{11} km.

Then we can see that at the end of its travel of 30,000,000" our vehicle will not find the Earth that remained back of d$s = s - s_T$ = 1,6346 × 10^{15} m. ds = 1,6346 × 10^{12} km.

To land shall be to do backward those 1,6346 × 10^{12} km. Never to think, obviously, to do them after the end of deceleration, it is restarting from still vehicle, because this would be the worst solution so by the needed time as by the needed energy. The best solution, instead, would be to make that operation at vehicle maximum speed, it is when $v \rightarrow c$, so to can exploit anew the relativistic space–time contraction. In this case, obviously, to run negative space, the vehicle must pass to negative space–time. But with v so near to C, the passing to negative space–time is so easy by energy viewpoint at it is delicate by the precision of maneuvers to accomplish. Think only that with $v = c$, in a zero elapsed time could be run an infinite space. The driving apparatuses should then be automatised in a way to can make pass the vehicle to negative space–time with its *v* speed enough inferior to *c* to not risk to make it arrive too far in a so short time to cannot control it. If we think to control the time at the 1/1000 second, it means we can control the running spaces, with $v \cong c$, inside the limits of 300 km, and this is enough for our purposes because for the landing, when a grater precision is needed, the vehicle will have more time and more fit means. In the given precision's order, we could choose, for the backward run, a *v*, in negative space–time, of 0,999,999C, which gives a $Fdt = (1 - 0,999,999) \, Fdt = 10^{-6}$, so to have: $s_T = 1,6346 \times 10^{12} \times 10^{-6}$; $s_T = 1,634,600$ km, which would be run in a time $t_c = s_T/c \, t_c = 5''448'666$.

Similar considerations must be done in the case on which the travel to accomplish should be very long so to have $s_T > s$. In such case, the correction run would not require to pass to negative space–time. With the same precision of time controlling and $Fdt = 0,000,001$, to reach the limit of the actual universe at 10^{23} km, or 10^{17} km, after relativity modification, it would take a time of $10^{17}/(10^5 \times 3) = 3,??? \times 10^{11}$ s $= 10,562, 69$ years. Such time would be too much for human beings, but it could be reduced, increasing the vehicle speed. So, with the universe limit at 10^{23} km from the Earth, the relation to obtain the vehicle speed to reach that limit in a year is $10^{23}(1 - v/c) = 31,557,600 \times v$; then: $10^{23} - 3,??? \times 10^{17} \, v = 31,557,600 \, v \, 10^{23} = 3,??? \times 10^{17} \, v$, as adding $31,557,600 \, v$ to $3,??? \times 10^{17} \, v$ does not modify the $3,??? \times 10^{17} \, v$.

Then $v \cong 300,000$. It is not exactly the light speed, but as much as possible near to it. This shall require to solve not easy problems of regulation of the engines' power of the vehicle, but we must think the purpose put ahead should not be of primary importance.

Again analogous reasoning must be alone for travels backward in time; for whose spaces correction the negative already reached space–time, gives negative spaces to add to the spaces due to accelerations and braking, as necessary, as from calculations. For spaces to detract from accelerated spaces, these must be run in positive space–time.

Chapter 22

Ballistics

From what was seen in the preceding chapter, the ballistic problem of space–time vehicles consists almost ever in making them reach the light speed, drive them at that speed for a very accurately controlled time to operate the space corrections normally necessary, and lastly brake them to make them return to zero speed at travel end.

We saw that the space correction operations, given they are made during the inertial motion and at the v ??? c, absorbs very little energy, so much that it may be neglected. It is needed instead to calculate the required energy for the two operations which practically absorb it completely: acceleration and braking.

Now, being our vehicles moved by rockets, the related dynamics founds on the third Newton law saying: 'To every action corresponds an equal and contrary reaction'. It is: ??? = ???, which in our case we can write:

(250) $Mv = mV$ where: M = ejected by the rocket mass at v speed, taken respectively to the vehicle, m = vehicle mass, V = vehicle speed respectively to the starting point ($V = 0$) at the exhausting of ejected M mass.

Our problem is then, primarily to determine M, when in the (250) are known so v as V, which are both equal to C because we want eject with matter fusion rockets, where, characteristically $v = c$, while our purpose is just to bring our vehicle to the speed $V = C$; but then brake it up to $V = 0$.

We also know m value at deceleration end, where it will be $m = Q$, while at travel starting it will be $m = Q + M$, where:

$Q = m_p + m_v$ (m_p = person's mass; m_v = empty vehicle mass)

$M = M_a + M_f$ ($M_a = Q$ + ejectable mass for acceleration; $M_f = Q$ + deceleration ejectable mass)

It is not the case to keep count of food mass and other consumption materials as toilette water and so on because they might be employed, after use, as engine propeller. They must be then considered as part of M at the ratio of 3 kg/day/person.

At this point we have enough elements at disposition to write the relations that lead to M. They are: for M_f:

$M_f = Q + ???\, dM_f$ (where dM_f is the differential of M_f corresponding time differential $dt = ta/2/n$). Then $dM_f = dE_s/C^2$ and $dE_I = (Q + dM_f)a \times ds$ (where dE_I = differential of necessary energy for a ds run space). At its time $ds = s_I - s_{(I-1)} = (½)at^2_I - (½)at^2_{(I-1)}$; $ds = (½)a(t^2_I - t^2_{(I-1)})$

Setting: $t_I = I_{dt}$ and $t_{(I-1)} = (I - 1)dt$ we have: $t^2_I - t^2_{(I-1)} = (Idt)^2 - [(I - 1)dt]^2 = (dt)^2[I^2 - (I - 1)^2]$ and: $[I^2 - (I - 1)^2] = 2(i - 0.5)$.

Then: $ds = a(dt)^2(I - 0.5)$ and $dE_I = (Q + dM_f)a^2(dt)^2(I - 0.5)$.

For M_a: $M_a = M_f + ???\, dM_a$ (with the same symbols meaning as over) Then:

$dE_I = (M_f + dM_a)a^2(dt)^2(I - 0.5)$ and: $dM_a = dE_I/c^2$.

At last: $M = M_a = M_f + \Sigma\, dMa$ (251)

M depends then by the value of two sums with a number theoretically infinite of addendums, but it is possible to obtain it practically at the convergence of calculations made at the varying of n towards as high values as possible.

It is then possible to try to dimension, even if coarsely, two typical space–time vehicles to have an idea of their factibility. We speak of a transport robot vehicle and of a person's transport one.

For robot's transport, to obtain the vehicle mass (propelling mass excluded) plus the robot in the 100 kg, the total mass of the vehicle at the start (propelling mass included) resulting from the (251) relation, will be: $M = 271{,}825$ kg. Pushing the robotisation to the point to have a vehicle able to supply itself with propelling matter (water) a time at destination, this will be able to undertake the return travel too in the dimension of 271,825 kg of mass.

The vehicle dimensioning requires, however, too the rocket-engine dimensioning, which, at its turn, depends on the maximum power it shall have to furnish. This is given by the relation: (252) $W_M = (MV)_M \times a$ (where $(MV)_M$ is the maximum value that can be reached by vehicle mass, by its speed (both instantaneous values)). In our case, with $a = 10{,}000$ m/s², $(MV)_M = 4{,}94613 \times 10^{10}$ kg·m/s and $W_M = 4{,}9461 \times 10^{14}$ kg·m/s $= 4.85 \times 10^{22}$ erg/s.

This power requires a matter ejection $M_{eM} = W_M/C^2 = 4{,}4955 \times 10^{-3}$ kg/s $M_{eM} = 5{,}4955$ g/s.

The engine-rocket dimensioning is obtained by considering its paraboloid ejector as a black body which emits radiation following the law:

$W = GT^4$ erg/s·cm² (253)

where $G = 5.66 - 10^{-5}$: Then: $W_M = 5{,}66 \times 10^{-5} \times T^4$ erg/s·cm² (254) From it: $T^4 = W_M/(5{,}66 \times 10^{-5})$ °K $T = 5{,}410{,}989$ °K.

Being this temperature corresponding to that of 1 cm² of surface black-body, thinking of making reflect such emission by a surface that can bear a 750 °K temperature (477 °C), the paraboloid surface must be: $S = T/750$ cm²; $S = 7{,}214$ cm², to which can correspond a rocket efflux section of: 2,400 cm², which gives a diameter φ = 55.3 cm.

This dimensioning for as much coarse it is, it's enough to guarantee the factibility of the vehicle for robots transport, with which, as seen, in only 8^h??? it is possible to run theoretically infinite time distances.

For persons transport vehicle the calculations are analogue and expressed by the (251) relation for what concerns the vehicle maximum mass at the start and by the relations (252), (253), and (254) for the rocket-engine dimensioning.

For the readers that would intend to verify the above shown calculations we give the BASIC program with which they have been executed at PC.

Space for Computer Prog.

The results are as follows:

M_f = 164.78, 164.86, 164.87, 164.87, 164.871
M_a = 271.52, 271.79, 271.81, 271.82, 271.825

A time chosen $n(N)$ = 30,000 as enough great value for the M_f and M_a determination, is possible to determine $(MV)_M$, as for that product maximum value it can be used, with enough accuracy, M = 164.871 kg and $V = C$.

Then we will have

$(MV)_M$ = 164.871 × 300,000,000;
$(MV)_M$ = 4.946 × 10^{10} and $W_M = (MC) \times 10,000$;
W_M = 4.946 × 10^{14} kg·m/s
W_M = 4.849 × 10^{22} erg/s.

These values bring to the dimensioning of the rocket as already done for robots transport.

For persons' transport vehicle, without entering newly in calculation details, we give here the BASIC program calculation results and the derived ones:

$M_a = M = 12,775,9217$ kg
$M_f = 7748.9891$ kg
$W_M = 2,3246 \times 10^{13}$ kg·m/s $= 2,279 \times 10^{21}$ erg/s
$M_{eM} = 0,258$ g/s
$T^4 = 4,0266 \times 10^{25}$ °K
$T = 2,519,045.72$ °K
$S = 3,358$ cm²
$\varphi = 37.75$ cm

Chapter 23
Epilogue

If travelling in time besides in space, will be possible with the over seen efficiency, there will have many important consequences for the humanity. Perhaps, the most important will be the multiplication of disposable space for the exponentially increasing human population. Up to now we have not pointed out to this possibility. Let's look at it now starting from the known paradox by which the backward in time travels incur in contradictions. It is said indeed that if time traveller, going back at the time when his grandfather was child and could cause his death, he would have not been generated, would have not been able to travel back in time, to meet the progenitor, to cause his death, and so on.

We can see that it is a serious objection that put at risk the believing of who, as us, think the time travels to be possible. Hawking, who believes as we do on this possibility, proposes the idea of 'cosmologic censure', by which the cosmos would take care to forbid the events that could generate contradictions with the already elapsed story. Other scientists (Hugh Everett) speak of parallel stories. This idea is more convincing because this does not require one to imagine which foundation to give and how should be the functioning of the cosmological censure. Indeed it seems logical to think that by the arrival from the future of a man to a space–time point of the past, he could live his life as all other persons of that time and, too inadvertently, cause the death of his predecessor, without changing the already passed history. So we are compelled to think that each time somebody or something reaches a given past time point in the universe, he or it causes the beginning of a new history, not superposed to the already passed one, but, just as stated by Heverett, parallel to that. Then, the principal history, the one we know, does not change. What happens is instead the beginning, every time, as branching from the main one, of a new, different and parallel to the main, history. Might already have had such events? Will already have gone men from the future to meet men of the past? Our history, the main one on which we live, has not transmitted us such facts, but the thing may not be excluded, especially if it happened in prehistoric times. But we might also not know it for a more special reason. It is that a new history, as branching from the main one cuts out us from its course. In other words, we can say that our history, already run, cannot be changed by different successive facts. These are part of another history and are the result of space–time travels. This means that at every instant of the main history (separated from passed and successive ones also only of the Plank's time (5.6×10^{-44} s) may begin a new and different universal history, and then on a different universe from that of the main history. Practically, each time a space–time travel will be accomplished, a new universe will be created, so that an infinity of them could be created depending on human will.

It will be good to put attention to the fact that the universes' diversity depends from their times. Then the dates shall have to be conveniently double; the unaltered one of their own time and that of the future time from which the vehicle started that changed the story creating the new universe. Another important consequence of space–time travels will be the possibility to make the lives of people multiply. Indeed each time a universe starts its existence branching out of another one, all the people living there begin a new life, even if already dead in the universe from which they branched.

Also the possibility to solve problems (of every kind) will have great help from space–time travels, because going to the future will be possible to see already realised projects that give much to think today. And about justice administration? Today, as ever before, there are crimes that can't be repaired. (Expiation is only a very poor substitute of reparation.) Think only of homicide. The detested and condemned criminal can only expiate the fault, while to make true justice would require to give anew the taken life. Now we know the space–time travels give us this chance. But the time travelling, too with only robots can help very much justice because gives the chance to detect the qualities going to take seeable documentation of the facts on the crimes scene. We can too see the importance of these undertakings for society's moralisation. Who would decide to commit crimes knowing he/she can be surely detected and condemned to reparation?

We can add, but without putting full stop, the scientific importance of space–time travels: it will be possible to visit all the planets and satellites of solar system, but also the planetary systems of our and other galaxies on its today, tomorrow, and too very far past. The limits of space and time will be indeed at human being's reach.

www.ingramcontent.com/pod-product-compliance
Lightning Source LLC
Chambersburg PA
CBHW020443220526
45464CB00002B/832